"Tom Nisbett has had a lifelong love of the mountains, and his stories give the readers a firsthand look at the impact of climate change and energy policy on this planet. Tom shares some of his most memorable experiences which allow the reader to better grasp humanity's impact on Earth's most beautiful and remote places. Tom's stories should be a wakeup call for each of us, before it is too late to save the mountains."

—BRYAN DAY, executive director, Little Rock Port Authority

"In *Moving Mountains*, Tom issues a compelling plea for wise stewardship of wilderness spaces by cultivating a love for sights unseen via the sharing of his own experiences and issuing an invitation to link arms in this quest by taking the accessible first steps of repeated exposure and firsthand observation. From mamas hiking local trails with their children to mountaineers scaling peaks across the globe, this book is sure to inspire!"

— NIKKI HUDSON, home educator and family adventure advocate

"Tom Nisbett, an experienced climber and explorer with an engaging, wise voice takes the reader on a series of treks to breathtaking destinations most of us will never visit, but, thanks to his words, the experiences come alive. His descriptions of exploring with others and learning together illustrate the spirit of humility and curiosity which mark each adventure."

—KELLY BEAN, author of *How To Be A Christian Without Going to Church: The Unofficial Guide To Alternative Forms of Christian Community*

"Tom's travelogue of climbing adventures gives the reader an insider's view on what the mountains can tell us about climate change. Tom's call for protecting our wilderness as stewardship of our planet for future generations is compelling and is rooted in his personal climbing experience as much as it is in his educational and spiritual worldviews.

—MATTHEW CLEVELAND, chief development officer, Arkansas Sheriffs' Youth Ranches

"Having known Tom for years while we were in Arkansas, I knew his heartbeat was to climb mountains. What I found in reading *Moving Mountains* was the pleasant surprise that the book's focus was on the effect of each mountain as it exposed the climate change that science keeps telling us is real. As an organic farmer, I see these effects daily and glad Tom and I are kindred spirits as stewards for the ones that come after us."

—BRUCE ROBERTS, organic farmer, Harpeth Moon Farm

Moving Mountains

Moving Mountains

Paying Attention to Weather and Climate

THOMAS NISBETT

RESOURCE *Publications* • Eugene, Oregon

MOVING MOUNTAINS
Paying Attention to Weather and Climate

Copyright © 2025 Thomas Nisbett. All rights reserved. Except for brief quotations in critical publications or reviews, no part of this book may be reproduced in any manner without prior written permission from the publisher. Write: Permissions, Wipf and Stock Publishers, 199 W. 8th Ave., Suite 3, Eugene, OR 97401.

Resource Publications
An Imprint of Wipf and Stock Publishers
199 W. 8th Ave., Suite 3
Eugene, OR 97401

www.wipfandstock.com

PAPERBACK ISBN: 979-8-3852-3418-9
HARDCOVER ISBN: 979-8-3852-3419-6
EBOOK ISBN: 979-8-3852-3420-2

03/06/25

For Lou Ann, Shawn, Candace, Mallary, and Savanah

All of you have been climbing companions of mine—in our family, on the mountain, in the celebrations of life, through the losses and grief, above the trials that would pull us down, out of the valleys and on to the summits, and I will say that together we have moved mountains. I couldn't have climbed so many without your companionship and your thoughts and prayers. It was said of Jesus that one day he climbed a mountain, and his climbing companions climbed with him. Only the faithful and the committed climb. "A faith that moves mountains is a faith that expands horizons, it does not bring us into a smaller world full of easy answers, but into a larger one where there is room for wonder." (Rich Mullins)

Be responsible for every living thing that moves on the Earth. Move mountains without doubt in your heart.

Contents

Title Description and Summary		ix
Preface		xi
1	Kilimanjaro, 2007 *Vanishing Glaciers*	1
2	Gannett Peak, Wyoming, 2008 *Oil and Gas Tax Subsidies*	15
3	Mt. Kenya, 2009 *Drought Impacts*	29
4	China, 2010 and 2012 *Air and Water Pollution*	43
5	Carrauntouhil, Ireland, 2012 *Rising Temperatures*	55
6	Colorado 14ers, 2013 *Violent Weather*	63
7	Coma Pedrosa, Andorra, 2014 *Water Resources and Snow Cover*	73
8	Chugach Peaks, Alaska, 2015 *Coastal Erosion and Pipeline Construction*	83
9	Snowdon, Scafell Pike, Ben Nevis, 2017 *Carbon Emissions*	95
10	Aconcagua, Argentina, 2018 *Temperatures and Glacial Melt*	107

Contents

11	Chicago Basin, Colorado, 2018 *Pine Beetles and Forest Fires*	121
12	Kosciuszko, Australia, 2019 *Drought, Forest Fires, and Coral Bleaching*	129
13	Antarctic Peninsula, Antarctica, 2023 *Rising Ocean Temperatures*	139
14	Mount Rainier, Washington, 2007 *More than Melting Glaciers*	147

Afterword	155
Endnotes	157
Climate Glossary and Definitions	161
Moving Mountains Playlist	163
Resources for the Reader	165

Title Description and Summary

The initial step for a soul to come to knowledge of God is contemplation of nature.
– Irenaeus

Environmental stewardship and wilderness protection are core values in outdoor programs and require a participant to *tell why natural resources are important to them and to the future of our country*. Today, environmental curriculum and nature-based lessons are taught in elementary and secondary schools. Beyond the intellectual learning in classrooms, experience in the wilderness and field research in nature is essential. Awareness is the work of the soul, and you must leave the cities, the industrial areas, the financial markets, and the rhetoric of the pundits to listen and to see for yourself.

The mountains particularly have much to teach us. Mountains are sacred places to people of all races, cultures, religions throughout the world. Mountains and glaciers are also a barometer of climate change and a bellwether of environmental trends. Mountains resemble leading indicators in the economic and political battle over climate change. Altering our destructive trajectory will be like moving mountains, *and* it will save them.

Preface

It has been a regular habit of mine to escape to the mountains. "Escape" seems the wrong verb. I was not breaking free from any confinement as much as I was running to the freedom of the hills. Though I would have liked to have lived in the mountains, I grew up only three hours from the Sangre de Cristo Mountain Range in northern New Mexico. I enjoyed hiking, climbing, camping, and snow skiing from a young age.

That practice continued into adult life and expanded into week-long backpacking trips in New Mexico and Colorado, the southern Rockies. *Escape became a curiosity to explore.* Exploring the mountains improves our understanding of our world and can change our perception of ourselves. My wilderness *escape* was rather an *entrance* to the natural world in the backcountry.

One year after our wedding my wife, Lou Ann, and I signed on for a three-week backpacking trip in the Grand Tetons Range in Wyoming. Because this adventure was organized and sponsored by Tarrant County College (TCJC back then), we earned and received college credit in Wilderness Literature, Geology, and Physical Education courses. After that trip, Lou Ann and I spent a winter ski season working at Angel Fire, New Mexico.

These outdoor experiences produced the desire in us to do adventure trips in the Big Bend National Park in Texas and the Weminuche Wilderness in Colorado. As we started a family, we did some hiking and camping trips in the Wheeler Peak and Pecos

Preface

Wilderness areas of New Mexico due to their proximity to our home in Texas. These trips were a great break in our busy lives.

Along the way, I earned bachelor, masters, and doctorate degrees in economics from three separate universities, earning concentrations in natural resources and economic development. I conducted research on water and agricultural resources in West Texas and community impact research of oil and gas production in northern New Mexico. I enjoyed conducting these studies but felt a calling to college teaching.

I had some awareness that modern industry was affecting the environment and the climate. However, I never connected my escape to the mountains which I loved with the economics and public policy courses that I taught and endorsed. And I never examined my biases and doubts about climate change. Nature has important lessons to teach if we listen. The mountains taught me to listen.

While I am an advocate for free enterprise and competitive markets, the market structure in the fossil fuel industry harms the Earth's climate and the government policies of subsidies and tax breaks for these companies perpetuate the problem. It required time away in the wilderness for me to really see the harm that greenhouse gases are doing.

Awareness is the work of the soul, and you must leave the cities, the industrial areas, and the financial markets to listen and to see beyond the air pollution and the toxic waste. The rhetoric of the pundits and the lobbyists cloud the issues. Our noisy and busy urban lives cause us to live alienated from the natural world and from ourselves. Mountains awaken us to what is real and essential.

To share my own understanding of the impacts of climate change on different countries and regions, I tell the stories of fourteen mountain climbing trips and travel adventures on seven continents that I made between 2007 and 2023. Unless you leave the comforts of home for the realities of the outdoors all around the world, you do not have firsthand knowledge.

Most climate change quotes are open share Wikipedia unless other citations are noted. Scientists always look for explanations

Preface

for what goes on in the natural world and test those explanations against evidence from the natural world. I do learn from the work of climate scientists because I am not one, and I share my experience as an aspiring mountaineer because I am one.

I have climbed hundreds of mountains and far surpassed Malcomb Gladwell's 10,000- hour rule for mastering a skill, though I consider that rule mostly incompatible with mountain climbing. Mountains move and change, expand and contract, grow taller and shrink sometimes. Weather and climate can completely alter the look of a familiar mountain as well as affect the mountaineer's judgment. Climbers constantly monitor the weather conditions and the changing climate dynamics. Even expert climbers die on the mountain.

I hope mountains will always exist. Mountains are sacred places to people of all races, cultures, and religions throughout the world. *From the Andes to the Himalayas, mountains have an extraordinary power to evoke a sense of the sacred.* (Bernbaum) Mountains and glaciers are a barometer of climate change and a bellwether indicator of environmental trends. Mountains are like leading indicators in the economic and political battle over climate change. Altering our destructive trajectory will be like moving mountains, and it will save them.

Thomas Nisbett
Ozark Highlands Base Camp

1

Kilimanjaro, 2007

Vanishing Glaciers

The greatest adventure is what lies ahead, unknown and uncertain.

Mountains are climbed by those who combine naivete and complexity. It's simple—just believe you can do it—and it's difficult because you never know what the mountain is going to throw at you. Mountains create their own weather and microclimates—pop-up thunderstorms with lightning and flashfloods and stalled snowstorms. The altitude affects your body. Everything in the mountains is moving.

Something like luck helped us in Tanzania in 2007. The best time to climb Kilimanjaro is during the dry season—from December to mid-March, and from late June to October. Just weeks before our January climb, the rain continued and made the experience a very wet one for some friends of ours. For us, the rains stopped in early January. *Luck is when preparation meets opportunity.*

Climbing Mount Kilimanjaro is one of those dream trips for hikers and climbers. Its name alone inspiring adventure and made popular worldwide by Ernest Hemingway's short story entitled, "The Snows of Kilimanjaro." It is one of the Seven Summits further popularized by the adventure and subsequent book with that title by Dick Bass, Frank Wells, and Rick Ridgeway about climbing the highest mountain on each of the seven continents.

My Kilimanjaro adventure started with a phone call from my son Shawn. I was naïve. I knew that Kilimanjaro was the highest mountain on the African continent. I had seen it from a Kenya Airways flight in June 2001. I had talked to family members and friends in Kenya about climbing it. Beyond that information, I had little knowledge about Kilimanjaro. I did learn that country boundaries were changed resulting in the mountain moving from Kenya to Tanzania. I knew nothing about its glaciers.

I didn't even know that it had its own international airport (JRO) out in the savannah between the two cities of Arusha and Moshi. When my son Shawn called me from Djibouti in East Africa where he was doing a land tour of duty as a U.S. Naval Officer, his idea to meet in Tanzania at JRO in January met no opposition from me. When are we going?

Kilimanjaro's glaciers are endangered. Hemingway's story in 1936 opens with a paragraph about the highest mountain in Africa whose western summit is called in Masai the "House of God." We are told that a frozen carcass of a leopard lies near the summit. No one can say why it is at that altitude nor why glaciers could exist at the equator.

> *The Furtwangler Glacier—named for one of the team members who climbed to the top in 1912— is located near the summit of Kilimanjaro. The glacier is a small remnant of an ice cap that once crowned the summit. Almost 85% of the ice cover melted from October 1912 to June 2011. At the current rate, most of the ice will disappear by 2040.*[1]

As Shawn and I, our team members Marc and Hettie from Belgium, and our National Park Guide, Makupa, began our climb in January 2007, my climate change awareness was activated by my mountaineering interest:

Kilimanjaro, 2007

Mountain climbing as a sport or human activity does not have a long history, especially compared to, say, sailing the high seas or fishing in the great rivers. Humans have of course crossed the Alps, or the Rockies, and other mountain ranges when they were going somewhere—migrating, escaping, or warring. But mountain climbing as an activity in and of itself can be traced back just over 200 hundred years, and perhaps only one hundred years or so as an actual sport.

The first ascent of Mont Blanc in France and Italy occurred in 1786 with the aim of making scientific observations and it would be more than three-quarters of a century before the Matterhorn between Switzerland and Italy was climbed in 1865. Peaks in the Alps began to be explored more by mountain climbers, creating the rich history of European climbers. The British Alpine Club was formed in 1857 and, over time, became the oldest mountain climbing association.

It would be the middle of the 20th century before mountains in the Himalayas were climbed—-Annapurna by the French team led by Maurice Herzog in 1950 and then Mt. Everest by Edmund Hillary and Tenzing Norgay in 1953. There are many expeditions in the history of the past 150 years and many sponsored climbers today. Why do we climb?

I did know the complexity. Shawn and I had been climbing mountains together since he was 8 or 9 years old. We had already hiked numerous 13,000 foot and 14,000-foot peaks and we observed weather patterns and frontal systems. We had dodged some wicked summer thunderstorms that popped up quickly in the Rocky Mountains and we had been chased from numerous summits by sudden lightning storms and unexpected hailstones. We learned to read the skies for changes in weather.

Shawn was a high school valedictorian and became an engineer and nuclear submarine officer with degrees from the U.S. Naval Academy and Old Dominion University. Our mountaineering skills were learned as well as acquired from experiences in different terrains and difficult route-finding challenges. Mastering weather and climate information comes with mountaineering.

Due to Kilimanjaro's proximity to the equator, the region does not experience the extremes of winter and summer weather, but rather dry and wet seasons. It is better to hike there when there is a lower possibility of precipitation. The primary issue is safety, as the risks associated with climbing increase significantly with foul weather.

The effects of rain, mud, snow, ice and cold can be very strenuous on the human body and reduce the chances of a successful summit. Most climbers travel a long distance and spend good money to make the Kilimanjaro trip. Constant awareness of the environment is a critical part of mountain education as is mental attitude. *I had little knowledge about glaciers and climate change.*

As we made our preparations and then our respective journeys to Tanzania, we began to read about climbing in East Africa. Shawn was traveling overland through Somalia and Ethiopia with a flight from Addis Ababa through Nairobi and into JRO. My flight took me from Houston through Amsterdam and down to JRO. Kilimanjaro and Mt. Kenya are the 1st and 2nd highest mountains in Africa, and both are multi-summit peaks. Both mountains boast their own ecosystems and create their own weather.

Kilimanjaro is about 200 miles south of the equator rising out of the East African plain. Because of its proximity to the equator and to the Indian Ocean as well as its 19,341 feet in height, Kilimanjaro possesses five major climatic zones—savannah and lower slopes; montane forest; heath and moorland; alpine desert; and alpine summit. But are there really glaciers in tropical Africa?

My reading included *Kilimanjaro & East Africa: A Climbing and Trekking Guide* by Cameron Burns and this note on climate change in his Preface to the Second Edition (2006):

> The worst things to see were the changes brought about by global warming. Comparing my 1997 and 2005 photos of Kili's glaciers showed that they had retreated considerably. Sadly, it's the airplanes and cars we use to get to the mountain that cause the greatest damage, and even when we are on it, we burn fossil fuels in the required stoves, further warming the planet.[2]

Kilimanjaro, 2007

In the *Global Warming Primer* by Jeffrey Bennett, the author presents Global Warming 1–2–3:

> Fact 1: *Carbon dioxide is a greenhouse gas that traps heat and makes a planet warmer than it would be otherwise.*
>
> Fact 2: *Human activity, especially the use of fossil fuels—coal, oil, and gas—is adding significantly more of this heat-trapping gas to Earth's atmosphere.*
>
> Fact 3: *We should expect the rising carbon dioxide to warm our planet, with the warming becoming more severe as we add more carbon dioxide.*[3]

I had paid little attention to this issue before our Kilimanjaro Climb. The glacial melt and decline would become real by Day Four on our ascent. Climates are always changing. Why, then, do we need to pay better attention?

There are five traditional trekking routes up Kilimanjaro—Marangu, Mweka, Umbwe, Machame, and Rongai—and there are massive cliff bands like the fabulous Breach Wall which presents technical climbers with harsh and difficult challenges. When we were there in 2007 the technical routes were closed due to the conditions created by the melting glaciers.

We chose to trek the Machame Route on Kilimanjaro. Day 1 we ascended about 4000 feet from the Machame Gate to the Machame Huts at 10,000 feet above sea level (asl). Day 2 we ascended to the Shira Hut and Shira Cave area near 12,600 feet. We were meeting people from all over the world as we hiked and camped—-Lawrence from Bermuda on his 3rd attempt; Alex from Wisconsin climbing her 1st mountain; two guys from France; and a group of four climbers from California.

Day 3 we ascended and descended terrain, did some rock climbing on Lava Tower (15,000 feet), and camped at 14,000 feet at the Barranco Hut. Day 4 was a long acclimatization hike of ups and downs to arrive at Karanga Camp in the Karanga River Valley at about 13,400 feet.

Late afternoon we hiked with our guide, Makupa, about a half mile up a northward-facing trail to acclimate further and to get a clear view of Kibo Peak. We were standing at about 14,200 feet and looking up to 18,400 feet toward Rebmann Glacier. We listened for an hour or more to the sound of the glaciers calving and crashing down the western side of the great mountain. We heard rocks and ice falling as the melting glaciers provided an unusual symphony that evening. Was this just a summer melt in the Southern Hemisphere or something else? Pay attention.

This memorable evening was just getting started! We had a mountaineer's dinner of tuna fish sauce over rice, cucumber soup, bread, oranges, and hot tea. Shawn and I stayed outside in the crisp air for an hour more as a billion stars came out above us and the evening clouds below us evaporated revealing the city lights of Moshi and Arusha in the distance. This was a dark sky area far from artificial light.

The skies were awesome with lightning in the west towards Mt. Meru and a mile above us Kibo Peak looming clear and strong. About 35 to 40 tents were pitched there at Karanga Camp representing a global village of 90–100 people. The Tanzanians sang traditional songs in their encampment and a 30 member Norwegian group gathered to share their choral music late into the night. *This was an extraordinary day!*

Johann Rebmann, a London-based Church Missionary Society member, landed on the east coast of Africa in 1846 and began to hear stories of an enormous mountain lying in the interior. On a second expedition in November 1848, Rebmann reached the village of Majame (Machame) and described the shape and snow on Kilimanjaro:

> *The western and higher peak is covered with snow throughout the year. By the Swahili at the coast, the mountain is known as Kilimanjaro (Mountain of Greatness) but the Wa-Jagga call it Kibo, from the snow with which it is perpetually capped.*[4]

The reality of snow on these two equatorial mountains (Kilimanjaro and Mt. Kenya) was not at first accepted in London, and learned discussions took place on the subject before the Royal Geographical Society wrote Halford John Mackinder, the explorer to make the first ascent of Mt. Kenya. The argument was not over whether snow-clad peaks could exist at the equator, as scholars were already aware of the high peaks of the Andes. The issue was whether these mountains in East Africa were tall enough to have snow.

We are having similar kinds of arguments today over the issue of global warming and climate change. What will mitigate the intense disputes? More evidence? More action? Despite the protestations of climate-change deniers, the politics and climate events are getting attention.

On Day 5 we slept until almost 7:00 am. It was another clear, beautiful morning at 13,400 feet. The sun rose over a lower eastern ridge and the temps warmed up faster. Most days by afternoon the clouds came rolling up the mountain from the warmer, moist air below us and engulfed the upper slopes, often blocking our view of the mountain above us.

It was an easy three-mile hike today to Barafu Hut at 15,200 feet. This was our High Camp positioned on a rocky, windy, bleak alpine desert with no vegetation. Barafu was our brief resting place and staging point for our summit attempt the next day. We ate some soup and tried to get some sleep.

We rose at midnight, layered up, re-packed, and left for the five-hour climb to Stella Point (18,915 ft.). Fueled by some hot tea and cookies, we climbed a very steep 30+ degree slope in the dark with our headlamps on, ascending almost 4000 feet. Stella Point is on the crater rim and the mountain here was covered with fresh snow. I have seen pictures in recent years when there was no snow on the rim trails. It was another one hour hike up the gentle rim trail to the summit at Uhuru Peak—19,341 feet. The nearby views on the summit included glaciers, ice needles, and the volcanic crater.

Mount Kilimanjaro is a dormant volcano with three volcanic cones: Kibo, Mawenzi, and Shira. It is the highest mountain in Africa AND the highest single free-standing mountain above sea level in the world. The first ascent of Kilimanjaro was made by Hans Meyer and Ludwig Purtscheller on October 6, 1889. In mountaineering and climbing, a first ascent (FA) is the first successful documented climb to the top of a mountain or the top of a particular climbing route.

At 6:25 am the sun rose on the roof of Africa, and we had clear views of Mawenzi to the east and Mt. Meru to the west. I even caught a glimpse of Mt. Kenya over 200 miles to the North. *I didn't know it at the time, but I would climb Mt. Kenya in 2009.*

We were not eager to leave despite temps near zero. We made quick calls to family on the satellite phone. I just had to call my dad and mom first. My wife Lou Ann never let me forget that fact—when I called her the battery on the sat phone died just as I told her, "We made it to the t . . . ." To this day she likes to tell people she didn't hear from me for three more days and wondered if I had fallen off the top! Usually, the mountaintop receives the best cell phone signal, but it has been my practice with cell phones not to call until we are down and off the mountain.

We snapped some team pictures and then began our descent and returned to Barafu Camp in two hours and twenty minutes. Six hours up and two hours back down to high camp! After resting for a few minutes, eating mushroom soup, and re-packing gear, we began another six-mile descent to Mweka Camp. It was a long, fast downhill hike over rock in the alpine desert and then through the moorlands zone.

Day 6 had started at midnight and so we had been climbing and hiking for 14 hours when we reached the Mweka Hut—a descent-only camp. We got some hot tea, mandazi bread, and fried bananas in the mess tent. I rested and wrote in my journal after we moved into our tent and Shawn visited with other climbers. He learned that Alex from Green Bay had summited it and was in camp. Fatigue finally set in. We covered approximately twelve miles and 14,200 vertical feet of gain and loss.

Two ideas related to *climbing and climate change* occupied my thoughts as I drifted off to sleep at 10,000 feet: (1) the effort is difficult and often requires adapting with brutal honesty—one of our team members made the tough decision to turn around on summit day. *Can we make the critical decisions we need to make*

Kilimanjaro, 2007

about climate? (2) often the journey is demanding and requires vigorous self-examination and real soul-searching—*Am I up to this challenge? Am I part of the climate problem?*

What do we learn about ourselves and human nature from venturing into the higher elevations and vertical edges of mountains? All wild places including mountains, deserts, and wilderness present us with danger and the opportunity to face our fears. We test our mountaineering skills and abilities against the daunting challenge of another mountain or a new route. The wild often contains unknowns.

But there are deeper and stronger powers in the mountains. One solid idea is that the mountains don't care about the human struggle. The mountain pays no attention to us, but we must pay attention to the mountain—the routes, the cracks, the lines. The indifferent mountain even draws or pulls, causing us to be *observant* to detail and aware of features we don't usually notice.

A related but different idea is that the mountain allows us to be there, and we climb its peaks by *cooperating* with it rather than fighting against it. The mountain ignores and the mountain consents. *Cooperation* is the purposeful interaction of two or more people, organizations, or systems. *Cooperation* is movement that pivots *observation* into action. *These two mountain realities are:* observation and cooperation.

Climbing solo up the summit rock of Granite Peak, Montana, provides a personal example of these two mountain realities. After crossing the snowbridge onto the south face of the summit pyramid, I looked for several chimneys observing the one I would climb to a prominent V-notch. The route then passes a jumble of ledges, cracks, gullies, cliffs, and chimneys. Good route-finding skills and using the mountain's features are necessary through this area. I got stymied by some Class 5 rock, but I could see the "keyhole" up above. I really worked with the mountain, cooperating to climb 15 feet up a chute using my arms, my legs, and my back. Past the keyhole onto the summit ridge, I scrambled to the summit at 12,799 feet.

Observing and Cooperating apply directly to climate change. The melting of Kilimanjaro's glaciers is undeniable and well-documented. It was an *observable* lesson with which to be confronted and, especially, on a successful and memorable climbing trip. I love the mountains and wilderness and will foster their importance and preservation though not a seasoned environmental activist or an ecology scientist by training. There is always an opportunity involved with change. *Cooperate.*

Our last day's hike from Mweka Hut down to the Machame Gate where we began our trek was easy and we were feeling the thrill of our accomplishment. I came to Kilimanjaro with some local knowledge and have gained a global understanding. This adventure and this climb were within the skillset of my mountaineering experience. The trip was familiar and previous trips to East Africa were preparation for Swahili culture. I traveled with my son and with great people.

I didn't expect to learn about melting glaciers and to return home with some harsh truths about climate change. But I will never forget the sound of glaciers calving and crashing down the mountain. Cooperation and acceptance of climate change is movement that directs observation into action. It is a fight for future generations not a fight against them. *It is required of good stewards that they be found trustworthy.*

My journal entries record lots of facts—meals, geography, plant life, weather conditions, customs, Swahili words, names. Saying goodbye to new climbing companions—Marc and Hettie, Alex (we gave her a yellow "livestrong" bracelet for summiting her first mountain), Craig and Michelle from LA, Lawrence, Makupa our guide, Jarrod our porter, and Anton our cook among others. Tips were paid to the Tanzanian guides and porters as we visited one last time at the Springlands Hotel.

Shawn and I sat by the Springlands pool for three hours that afternoon soaking up the equatorial heat. We had a Kilimanjaro premium lager to celebrate! We were able to have dinner with a local friend. Remmy's family owns the Hotel, and she comes from Toronto to help her sister in the busy season. Her father was Somalian, and her mother was Tanzanian.

Kilimanjaro, 2007

Remmy took us to a local restaurant and bar in Moshi on the unpaved back streets and ordered different meats including goat ribs with hot dipping salsa and we ate out front, sort of like a sidewalk café. The bill was $11 for the three of us! Remmy showed us around Moshi which was a real cultural experience. *Great people make a great difference, and they are usually the change agents!*

I love weather apps, the Weather Channel, and studying the weather. It is my portal into climate change and topics like global warming. Mountaineers have a rich supply of weather information available to them today both before making an expedition and during the climb as well.

All the weather information must be applied and put into action to be relevant. I recorded mountaineering notes, environmental remarks, people impressions, and weather *observations* in my climbing journal on Kilimanjaro. Here are some on weather:

1. We enjoyed bluebird weather all seven days on the mountain.
2. We had very little rain (even in the rainforest zone) like they had just weeks before.
3. I used SPF-45 sunscreen every day despite the morning cloud cover.
4. The summit day dawned clear and cold with light winds on top.

My environmental *observations* written on the plane trip home:

1. Kilimanjaro's environment and ecology are changing.
2. The glaciers are melting and shrinking due evidently to global warming, solar activity, or longer weather cycles.
3. The mountain was very dirty with trash and human waste.
4. The human impact is huge with 100,000 climbers annually, and the sheer number of hikers, guides, and porters.
5. Most water resources for the villages and cities below come from the rainforest, not just the snow up higher. But the rainforests and the snow melt are being altered.

Moving Mountains

Let me recap: I was naïve and knew very little about climate change in 2007 when I wrote these comments. Global travel and adventure always provide learning opportunities. Adventures and opportunities tend to find us, interrupting our expectations. I love the mountains and the benefits they confer on our material well-being and on our spiritual health. My spiritual mentor repeated many times: *Whatever gets your attention gets you.*

Preserving wilderness and wild places for future generations has a new dimension today: paying attention to weather and to the changing climate. *Be responsible for fish in the sea and birds in the air, for every living thing that moves on the face of Earth.*

2

Gannett Peak, Wyoming, 2008

Oil and Gas Tax Subsidies

Life is either a great adventure or nothing.
—Helen Keller

The Teapot Dome scandal (1921–23) involved bribes in the President Harding administration by private oil companies for land leases in Wyoming. The U.S. Secretary of the Interior, Albert Bacon Fall, went to prison based on this case of government corruption and bribery. Wyoming is big and its landscapes are inspiring but look closer and the state has a long record of corporate neglect and environmental issues with social impacts.

Even with several trips to Yellowstone and hiking adventures in and around the Tetons for multiple weeks at a time, I was unaware of Wyoming's long history of oil and gas drilling and pipeline infrastructure. I went to the mountains there to climb.

In his book *Highpoints of the United States: A Guide to the Fifty State Summits*, Don Holmes writes:

Moving Mountains

> *Whether your goal is to reach the highpoint of your state or the highpoint of each of the fifty states, it is a goal shared by many across the U.S. You may be an experienced mountaineer accepting the challenges of Mount McKinley (Denali) or Gannett Peak (WY), whatever your level of skill or interest, the highpoints of the U.S. offer a diversity of experience.[5]*

A good percentage of those who climb mountains grew up on or near great mountain ranges or moved there as soon as they could. Our lead guide on Kilimanjaro, Makupa, grew up in the village of Machame on the southwest side of the mountain. It is a humble living whether growing tea or repairing old vehicles. Makupa could make a better income by guiding climbers. Although mountain climbing began as a pursuit of the wealthy and their agents like with golf or lawn tennis, mountaineering has become a middle-class pastime and hobby of common folks.

My dad, Joe Nisbett, lived near the Sangre de Cristo Mountain Range and ran across the Holmes book and the "50 Peaks Challenge" in early 1992. We decided quickly that it was a challenge we wanted to take on. Dad turned 74 in 1992, and we would go on to hike 17 of the 50 highpoints together. Perhaps the most exciting one would be Mt. Elbert, which is Colorado's highest at 14,433 feet. We summited it on July 7th, 1999, when dad was 81 years old. Dad and I knew we needed a special three-day plan if he was going to reach the top. Dad lived to the age of 96 and he followed our highpoint adventures with great interest. *Gannett Peak would be my 49th State Highpoint.*

Shawn and I were committed to attempting Gannett Peak in 2008 and we invited my climbing companion Kent Barnard from Bakersfield, CA, to join us. Gannett Peak is in the Wind River Range on the border of Bridger Wilderness and the Fitzpatrick Wilderness and is the most glaciated mountain in the Rocky Mountains. At least five glaciers surround the Peak—Dinwoody, Gannett, Gooseneck, Minor, and Mammoth—with more glaciers on route.

Gannett Peak, Wyoming, 2008

The primary route is 40.4 miles round-trip on trails, cross-country, and on snow and glaciers. Gannett Peak is in the heart of a remote and rugged wilderness. Because of this remoteness, elevation, and extreme weather, it is often considered by mountaineers to be one of the most difficult U.S. state highpoints to reach. The Dinwoody Glacier and Gannett Glacier have been the subject of numerous studies in recent decades. All the glaciers are melting. *What is the impact on the mountain ecosystem I wondered? How does the impact affect the people who are reliant upon the water source?*

We were serious about our physical training for climbing Gannett Peak, and that had something to do with its glaciers. One week before flying to Salt Lake City for the drive to Pinedale, WY, my daughter Candace, carrying her two-year-old son, joined me for a climb to the top of Wheeler Peak (13,161 ft.) in New Mexico.

We then drove straight down to Dalhart to run the XIT 5K race that my dad had organized and run since the late 1970s. Dalhart's elevation is roughly 4000 feet asl and this was the 31st running of the annual race. My parents moved there from Amarillo, Texas, when a job offer from the District Judge, Harry Schultz, required that they live in Dalhart where the District Office was located. Dalhart folks loved their ranching heritage.

The XIT Ranch was the largest ranch in the world at one time with over three million acres and 6000 miles of fence. I grew up around many farmers and ranchers and the businesses dependent upon agriculture. There were very large farms and ranches, and none were today's corporate entities. These Panhandle people were good stewards of the land, good members of the community, and people of faith.

My maternal grandfather was a dryland farmer most of his adult life in Oldham County about 65 miles south of Dalhart. In the book, *The Worst Hard Time,* author Timothy Egan calls the people of the Texas Panhandle and High Plains "tomorrow people—spartan and resilient." *We were hoping to show that resilience on the Gannett climb.*

Moving Mountains

Five western state highpoints are rated Class 4 on the HP difficulty scale—Alaska, Washington, Oregon, Montana, and Wyoming. These five peaks—Denali, Rainier, Hood, Granite, and Gannett—involve more difficult climbing with considerable exposure or travel on glaciers or snowfields where ropes are required.

The sentence that really caught my attention in one of guidebooks regarding Gannett Peak was this one: *Due to sudden storms with strong winds and subzero wind chill, climbers should not expect this summit to be reached on the first attempt.*

From the town of Pinedale, Wyoming, we drove to the Elkhart Park Ranger Station and Parking Lot for access to the Pole Creek Trailhead on a bluebird day. This primary route begins at 9500 feet and approaches Gannett Peak from the south along the Pole Creek Trail, the Seneca Lake Trail, the Indian Pass Trail, and the Titcomb Basin Trail.

On Day 1 we hiked this great trail through spectacular wilderness for about 12 miles to Island Lake, ascending to 10,400 feet. *One rule of thumb in mountain hiking on trails estimates one hour per 1000 feet of ascent or one mile per hour.* With less than 1000 feet of gain over 12 miles, our team of three covered the distance in just over seven hours. Not bad on our first day!

It was common for many years to see very few people in the backcountry in U.S. wilderness areas. People flock to the popular national parks like Yosemite, Yellowstone, or the Grand Canyon but most didn't hike into the mountains for an extended camping trip. With the growth of camping and big retailers like Bass Pro Shops, Academy, Dick's Sporting Goods, Cabela's and many others, *that has changed!*

Scenes of hundreds and more hikers like I saw on my annual Colorado 14ers trip in September 2020 are changing the human impact on pristine areas. The 2020 coronavirus pandemic has fueled this move outdoors and will likely continue drawing people to our national parks and forests. *Pay attention.*

The State of Wyoming faced numerous environmental issues from human impact beyond melting glaciers: oil and gas development characterized by boom-bust cycles; air, water, and soil pollution; overgrazing by livestock on leased land; wildlife habitat destruction impacted by the Pinedale Anticline Project Area and the Jonah Gas Field.

External costs accompany economic development and the growth of local and state economies. As an economist I know that there are many costs (externalities) associated with fossil fuels not reflected in the price of energy but are passed along to society like pollution, health impacts, and subsidies or tax-write-offs for fossil fuel companies.

Jeffrey Bennett makes this provocative statement: *I would argue that our current energy economy is essentially a form of socialism. By a "socialized cost" of energy, I mean any cost that is real but that is borne by society rather than by individual energy users.*[6]

In 2008 in Wyoming, I primarily saw the visible benefit to the local economy with a shiny new community center in Pinedale and the temporary job gain. Hiking in the protected wilderness usually spares us from these social and economic issues for a while. *When camping beside glistening alpine lakes, hiking below bluebird skies, and seeing practically no people, wilderness replenishes the soul. God created the alpine meadows, and the quiet waters and God restores my soul. We need protected wilderness!*

On Day 2, enjoying great August weather, we hiked six miles from Island Lake to the base of Bonney Pass at 10,800 feet just past Upper Titcomb Lake. Camp 2 would serve as our High Camp for our first attempt at Gannett's summit. Shawn and I left our climbing companion, Kent, near Bonney Pass at his request, where he would wait for our return on summit day.

Day 3 started early (5:15 am) as we crossed Bonney Pass and proceeded onto and across Dinwoody Glacier to Gooseneck Pinnacle Ridge passing the bergschrunds and crevasses up the summit ridge crossing the snowfield to the summit of Gannett Peak (13,804 ft.) at 2:30 pm. This section of the route was about two miles from Bonney Pass to the summit which we covered in about two hours.

In 2008 all this route was snow-covered. Like most state highpoints, there is a USGS benchmark and a register at the top but there was too much snow to find it. This was a long, tough day that included six miles roundtrip, 6000 feet vertical, a vertical snow-wall with kick-steps, and snow travel with crampons. Our day stretched into 17 hours arriving back to High Camp at 10:45 pm in the dark. Finding my tent, I just collapsed on my sleeping bag and fell asleep instantly. *This was a good exhausting day.*

Gannett Peak was officially named during 1906, in honor of Henry Gannett, Chief Geographer for the United States Geological Survey as well as one of the founding members of the National Geographic Society. Gannett Peak is the highest peak in the state of Wyoming and was first climbed by Arthur Tate and Floyd Stahlnaker in 1922.

As noted, the weather in the high country of Wyoming can be stormy and wintry even in mid-summer. The summit has a tundra climate, with short, chilly summers and very long, very cold winters. The wettest month is April, and the precipitation (mostly snow) occurs year-round. We had a massive high-pressure system and, therefore, spectacular weather for our eight-day backpacking trek.

I never take great weather for granted and almost always pack the right layers for all conditions. Weather forecasts are not always reliable, and mountain ranges generate meteorological anomalies. Human activity also alters the alpine environment.

The *observable* lessons of recent trips, including this one in the Wind River Range of Wyoming, were learning more about human impacts on the mountain environment and the hidden costs to the economy. The lessons are harsh on the one hand because I prefer freedom and the solutions that markets and entrepreneurs bring to economic problems.

No one likes more government regulation, taxation, and bureaucracy represented by policy solutions. These lessons are harsh because we all must examine our thinking and consider our actions. Society and the greater good can be served by reasonable regulations. Every time I take a commercial flight, I'm glad there is an FAA regulating aviation safety.

Gannett Peak, Wyoming, 2008

There are now many templates and exercises for people to measure their carbon footprint. *I've done that.* It can be educational for an open-minded person. Individuals and even families can lower their home thermostats, bike to work, drive an electric vehicle or dry their laundry on a clothesline. *I've done some of that.*

The bottom line of all these studies and individual actions still doesn't reduce the carbon (greenhouse gases) in our atmosphere. *The efforts we need to reverse this adverse impact involve bigger actions and larger polluters.* Economists don't often agree on many things with some on the liberal side and some on the conservative side. Even when facts and evidence support an economic theory, economists seem to enjoy disagreeing. Most economists generally support this policy recommendation, as stated by climate scientist Jeffrey Bennett:

> *The obvious way to institute a true free market for energy is to build the currently socialized costs into the prices paid in the marketplace. Economists across the political spectrum agree that there's one surefire way to do this: institute a carbon tax that accounts for the socialized costs, so that market prices (with the carbon tax included) will reflect the true costs of fossil fuels.*[7]

This policy discussion boils down to how high the tax should be, for what the tax revenue should be used, and who really pays it: producer or consumer? Do we phase out fossil fuels at a faster pace? Can alternative energy sources like solar, wind, and geothermal really generate enough power to replace non-renewables? What is the role of nuclear power going forward?

These are tough questions for us with no easy answers. I became an Eagle Scout and vowed to protect and conserve our country's natural resources—its soil and water, its woods and fields, its wildlife—long before we knew much about the harm being done by greenhouse gases. I earned a doctorate in economics emphasizing natural resources and regional development before understanding the carbon impact. *What lessons does my old friend the wilderness want to reveal?*

Moving Mountains

Approaching and summiting Gannett Peak in three days gave us five more days to explore and enjoy the backcountry in the Wind River Range. Day 4 we decided to pack up our gear and hike 8 miles from the Upper Titcomb Basin to Seneca Lake just south of the juncture of the Indian Pass Trail and the Highline Trail. The alpine lakes in this wilderness attract fishermen for the six species of trout in the streams and lakes. I tried to fly fish some, but I did more relaxing than catching!

I once read an article that identified what the author called the "silent sports." They are mountaineering, fly fishing, kayaking, and trail running. The silent sports produce a "hard-won grace" according to the writer. I've thought a lot about that since I enjoy these sports. Each can be done alone and often are. No crowds are cheering you on.

These sports are often done non-competitively and no medals are awarded. The "hard-won grace" is something like honor or favor held inside. External reward or praise is absent. *When we internalize a truth, a principle, or a standard, we come to embody it.* The "reward" for wilderness preservation will benefit future generations .

Here is an example from mountaineering. Mountaineers seek the uncharted way, the trail less traveled, and a summit to stand on. We recognize that the wilderness we seek is a resource that we must protect. This principle internalized is the mountaineering ethic:

> *As mountaineers we do our part to protect and preserve the wild country, we explore by applying Leave No Trace principles, using good judgment, and educating others. When we enter the backcountry, we are active stewards and contribute to the lasting protection of wild resources for ourselves and for future generations.*[8]

On Day 5 we moved our camp again and hiked about six miles down from Seneca Lake to Miller Lake. My journal entries indicate a week of brilliant weather and "bluebird" days. When a large high-pressure system decides to just settle in over the Rocky Mountains for a week or so, you have chosen the right week to climb and camp. That's what we got—and it was probably a combination of planning and luck!

Beautiful and even sunny days in August at elevations above 10,000 feet are usually accompanied by cool or cold nighttime temps. One night while camping at Miller Lake I woke up at about 3:00 am and needed to go outside the tent to relieve myself. It was a cold, clear night. It was dark. So, I stayed in my sleeping bag wide awake for a few minutes longer.

As I reflected on my 49th state highpoint (on our first attempt) and all the travel, the existential question formed in my mind—*How did I get here?* I knew the answer of course. The larger question followed—*How do you get to where you are going?* Like a divine download or moment of inspiration, the complete answer dropped into my head! *There are Three Secrets to getting where you're going:*

1. *Tell me where you're going.* Dad always said, "If you are failing to plan then you are planning to fail. I had laid out a written plan for the state highpoints. As a self-employed organizational consultant, I combined business trips with state highpoints. I combined high points in adjoining states. Write out your goals and destinations!

2. *Tell me who is going with you.* I've made some solo climbs, and they have value for testing my skills. Most hiking and climbing experiences are better shared with quality climbing companions. At least ten friends and family members have joined me for one or more highpoints. On mountains like Rainier and Hood I roped together with some total strangers. Most achievements in life require a team effort. No one gets to the top alone.

3. *Tell me what resources you need to get there.* Make a budget. Count the cost. Get the right gear. Set aside funds regularly for future trips. But, never let money be the reason you pursue your plans and dreams or don't. Identify the resources you will need, work to secure them, and then expect them to be there.

All of that came to my mind in one minute lying in my tent in the mountains. For a dozen years now, I have used the *Three Secrets* with individuals, organizations, staff members, and nonprofit boards to inspire action plans. Written action plans incorporate *observation* and *cooperation*.

We will need a plan for addressing climate change and transitioning from a fossil fuels economy: *Where are you going? Who is going with you? What resources will you need to get there?*

On Day 6 our friend Kent was still resting and reading at our Miller Lake camp. Shawn and I decided to hike the three miles out to the Elkhart Parking Lot and drive into Pinedale for a much-needed cheeseburger! In town we spotted the new community center built with the help of oil and gas funds from energy development in the area. The community center folks were kind enough to let us use the restrooms and showers to clean up before finding the best burger place in Pinedale.

It has been our practice to ask locals where to get the best burger when we come out of the backcountry. A week or more of dehydrated foods, trail mix, and instant coffee gets old. Over the years I have kept a list of my favorite burger joints near the mountains that may or may not still be there:

- *PastimeBar—Leadville, CO*
- *Burger Barn—Pinedale, WY*
- *Grizzly Bar—Roscoe, MT*
- *K's Dairy Delite—Buena Vista, CO*
- *Huckleberry Inn—Government Camp, OR*
- *Red Rock Café—Wall, SD (try the buffalo burger)*
- *Stray Dog—Taos Ski Valley, NM*

With full bellies and renewed energy, Shawn and I hiked the three miles back to Miller Lake in just over an hour. We had never left camp and come back like this before on any of our climbing trips. We slept well that night!

Gannett Peak, Wyoming, 2008

Day 7 was spent relaxing, flyfishing in the lake, and exploring the area. After dinner that evening, we were permitted to build a large campfire. It burned late into the night while we talked about many topics and issues. Walking and camping together allows people to really build thick relationships (as opposed to thin ones today due to busyness and social media). My climbing companions are some of my closest and best friends and family members. I trust them because we've built trust one step at a time.

When I was pursuing most of the state highpoints in the Western U.S., my climbing friend Kent drove from Bakersfield and would make the hikes and climbs with me. He usually drove to the airport in his Toyota Prius and met me. I remember getting sixty mpg numerous times as we drove across vast deserts and mountain ranges. We did seven western state highpoints with that plan—California, Nevada, Utah, Idaho, Oregon, Washington, and Wyoming.

I took a flight to Las Vegas, NV, and we combined Mt. Whitney (CA) and Boundary Peak (NV). On another occasion I flew into Salt Lake City, UT, and we combined King's Peak (UT) and Borah Peak (ID). The Wyoming trip described in this chapter was a single state highpoint. When I flew into Seattle, Kent picked me up and we combined Mt. Rainier (WA) and Mt. Hood (OR).

We were so fortunate to enjoy each other's friendship and so lucky to climb six of the seven mountains on the first attempt. We were blasted by a June 2007 snowstorm on our first attempt at Mt. Hood and had to turn back at the top of the Palmer Glacier in a whiteout. I went back alone in May 2008 and summited Mt. Hood with a team from Timberline Mountain Guides. *Kent could not go with me, but he was there with me in spirit.*

On Day 8 in the Wyoming backcountry the three of us broke camp, packed our gear, and hiked the three miles out to the Elkhart lot. This adventure had been a special one and our only regret was that Kent did not feel strong enough on summit day. Some days on the mountain are like that. You can dig deeper and even push the envelope, but you must know your limits. I think this is a Jim Whitaker quote: *Just because you love the mountains doesn't mean the mountains love you.*

Moving Mountains

Climber Ed Viesturs was named National Geographic's Adventurer of the Year in its December 2005 issue. Ed made this statement that year: *I've learned in climbing that you don't 'conquer' anything. Mountains are not conquered and should be treated with respect and humility. If we take what the mountain gives, have patience and desire, and are prepared, then the mountains will permit us to reach their highest peaks. I believe a lot of things are like that in life.*

Lots of things are like that. The Bureau of Land Management (BLM) is still going to sell oil and gas leases to exploration companies in the transition from a fossil fuel-based economy to more sustainable and cleaner energy sources. External costs of greenhouse gases and carbon dioxide emissions will still occur as we move toward carbon reduction and cleaner environments worldwide. That effort requires companies to combine immediate private action with longer-term strategies in the marketplace. We either buy or don't buy their products. Vote with your dollars.

3

Mt. Kenya, 2009

Drought Impacts

One's destination is never a place, but a new way of seeing things. —Henry Miller

In September 2009 I was traveling in Kenya, and I spent the first week in Mombasa on the coast bordering the Indian Ocean. A Texas church had a work team that was building a permanent church structure, and a teaching team that was sponsoring a pastor training conference. I was able to travel with them from the DFW airport in Texas through Amsterdam connecting to Nairobi with a short flight to Mombasa.

It was a great week that included construction work, a photo game safari in Tsavo National Park, and eating at different restaurants in this diverse city: Ethiopian, Indian, and Swahili. At the end of the first week, five of us took a flight back to Nairobi where four of the Texas guys were catching their return flight home and I was preparing to travel north to climb Mt. Kenya.

I met up with a driver named John with Mount Kenya Guides (MKG) and he drove me from the Nairobi airport into the heart of the city to the Kenya Comfort Inn. I checked into the hotel, paid John the $630 fee for the Mt. Kenya climb plus this lodging and the airport transfer. The hotel was in the center of downtown and I was advised to have my dinner and breakfast there and not to go out and explore.

Hotel guests and others partied late into the night, and I could hear gunshots or fireworks on the streets below at times. It must have been a special occasion! People returned to their rooms at all hours of the night and one or two guests tried to get their key to open my door. I didn't sleep much that night.

I do like traveling to different countries and growing in my understanding of other cultures. I once stood in line to visit with author Stephen Covey following his presentation to get his autograph in my copy of *Principle-Centered Leadership*. As I moved closer to him signing books, I noticed that he looked each person in the eyes and wrote something unique to each person above his signature.

When it was my turn, he studied my eyes and then in my copy he wrote "Kaizen Tom" and "Never give up." The Japanese word *Kaizen* denotes continual improvement or lifelong learning. He read me like a book! I am an avid reader (inherited from my parents) and hope to always be a lifelong learner. I began to read and to learn about climate change.

Climate change in Kenya is increasingly impacting Kenyans, their livelihoods, and the environment. Climate change has led to more frequent extreme weather events like droughts which last longer than usual, irregular rainfall, flooding, and increasing temperatures. The effects of these climatic changes have made already existing challenges with water resources, food security, and economic growth even more difficult.

Harvests and agricultural production which account for one-third of GDP in Kenya are also at risk. The increased temperatures, rainfall variability, and strong winds associated with tropical cyclones have combined to create favorable conditions for the breeding and migration of pests. For example, in early 2020 some parts of

Mt. Kenya, 2009

Kenya and other East African countries faced massive swarms of locusts.[9]

As we left Nairobi following my mostly sleepless night, another MKG driver was transporting me and my gear on a three-hour drive to Nanyuki. I soon began to see terrible drought conditions and dead livestock along the rural highway north. I asked William the driver about conditions, and he confirmed the severely dry weather adversely affecting the people and agricultural production. Neither of us mentioned "global warming," but no one who cares will fail to *observe* the changing climate patterns in their own regional area.

Kenya is a democratic country better off economically than many African countries. Nairobi and Mombasa are bustling and growing cities. Kenya is still classified as a developing country in contrast to a developed country. Climate change is potentially damaging to all agriculture production, particularly subsistence farming and harsh on people living at poverty levels.

Nanyuki town is located on the northwest side of Mount Kenya NP with close access to the Sirimon Gate. Arriving in Nanyuki, we gathered at the Ibis Hotel for a meeting with one of the MKG guides and getting some morning tea. Some climbers who were going to be on our team and climb Batian or Nelion—the technical summits of Mt. Kenya—were delayed and the lead guide Robin was going back to Nairobi to collect them. After some delay and decisions at the hotel, my guide was going to be Alex who I would meet at the entrance gate.

Robin drove me the 20–25 kilometers there and introduced me to Alex, Daniel the cook, and Peter who would porter the gear. Our goal would be to climb to Point Lenana at 16,355 feet and back over the next four days. I've always wanted to return and climb one of the technical routes which rise to 17,058 feet as we originally planned.

Since Robin drove me to the Sirimon Gate, I had the chance to ask him a favor. I told him about my client in Mombasa and he knew of the excellent eye care provided by the Lighthouse for Christ Eye Centre there. One of our fundraising events in the U.S.

was the annual charity golf tournament in Dallas, Texas, dubbed the "Swing for Sight."

I am not a golfer and because I climb mountains, I got the LFC board's approval to promote a "Climb for Sight" pledge campaign for my Mt. Kenya climb. Robin loved the idea and told the guide to allow me to summit first and get some pictures. We raised over $5000 in donations for LFC. Inside the Gate entrance, I ate my sack lunch provided by MKG, and we headed up the Sirimon Trail at 12:45 pm.

Sirimon Trail is an old roadbed, and some people choose to drive this section but hiking it to acclimate is recommended. This was an easy hike of 9–10 kilometers and about 600–700 meters vertical gain to Camp 1—the Judmaier Camp and the Old Moses Bunkhouse. (10,827 ft.) We covered the distance of roughly six miles in two hours and 15 minutes.

The Old Moses Camp has huts for cooking and eating, sleeping (about 50 beds), and showers. We quickly began talking with other hikers and getting acquainted. My room had a young doctor, Fergus, from the UK, a newly married Jewish couple from Israel, a couple from the Czech Republic, two more UK climbers, and a guy named Curt born in Minnesota but from Colorado. Curt had lived in Nairobi for 4 years and out of the U.S. for 18 years working for an NGO. After dinner the temps at this altitude were cooler and I decided it was time to get to sleep. *Day 1 had been a great start!*

Mt. Kenya is a mountaineers' mountain. It resembles some of the legendary peaks of Patagonia, the Matterhorn, and certain mountains in the Canadian Rockies. Mount Kenya's peaks lie just 17 kilometers south of the equator and boast the best ice climbs in Africa. The mountain is an extinct volcano in the Central Highlands with several distinct climate zones: the cultivated slopes around the national park boundaries, the bamboo forests, the high moorlands around the main peaks, and the alpine desert zone.

Mt. Kenya National Park was created in 1949 and in 1977 was designated a World Heritage Site by UNESCO for being *"one of the most impressive landscapes of Eastern Africa with its rugged glacier-clad summits, Afro-alpine moorlands, and diverse forests."*

Mt. Kenya, 2009

The Kikuyu, who make up most of Kenya's modern-day population, called the mountain Kirinyaga, which, roughly translated, means "white or bright mountain." They believed the mountain to be the home of their god Ngai, and out of respect, some still build their houses so that the doors face Mt. Kenya.[10]

We follow the footsteps and stand on the shoulders of those who have come before us.

I woke up on Day 2 well-rested at about 5:45 before wake-up calls were scheduled at 6:00 am. These guides run on schedule! I packed up with plenty of time left for a big breakfast at 6:30 pm. I had porridge, fruit, tea, toast, scrambled eggs, pancakes and jam. Alex and I left at 7:10 am and hiked up the Mackinder Valley Route past a creek bed of the Ontulili River.

The trail traverses along the northern slopes gradually turning south, cresting a ridge, and then descending a valley, crossing Liki North River, and making the ridge above the Mackinder Valley. In my experience the local climbing guides knew some English and common expressions for interacting, but longer conversations including their interests and background were hampered by the language barrier. I wished I knew more about Alex's story.

We hiked up to Shipton's Camp, a total of about 8.5 miles, in about 5.5 hours. Arriving at Camp 2 at 12:40 pm, we had ascended another 1000 feet to 13,800 feet. The skies had grown overcast by mid-morning and were spitting snow pellets at us as we hiked into Camp. Shipton's Camp sits at the base of all three peaks—Lenana Point, Nelion, and Batian—all named by Halford Mackinder after legendary Masai *laibons* or medicine men.

Mackinder and two climbers named Ollier and Brocherel made the first ascent (FA) of Mt. Kenya on September 13, 1899, although the people groups of Kenya believed God (Ngai in Kikuyu) resided on this mountain. Local climbers regularly ascended the peaks to perform spiritual rites. It is important to all the ethnic communities living around it—Kikuyu, Ameru, Embu, and Maasai. They all consider the mountain as an integral aspect of their culture.

Lunch at Shipton's Camp included bologna sandwiches, hard-boiled eggs, hot noodles, and tea. I laid down for a thirty-minute rest between 2:30 and 3:00 pm. After that, snow was falling harder, so I hiked up above 14,000 feet to acclimate and familiarize myself with tomorrow's summit trail. We will have an early start at 3:00 in the morning.

It was raining back at camp while more snow was accumulating higher up. I spent the late afternoon at the bunkhouse visiting with Peter and Amanda who are young doctors from Australia working at Homa Bay Hospital on Lake Victoria and welcomed Alawn and Adi from Israel. The five of us played Uno for an hour before dinner and rocked the place with laughter.

We started to learn Mini-Bridge that I had never played, but then dinner was served. We had mushroom soup, bread, fruit, and spaghetti. MKG encouraged us to drink a lot of tea to hydrate for the summit climb. We would rise at 2:30 am, layer up, have tea and biscuits, and hit the trail by 3:00 am. I headed for bed at 8:00 pm. *This was another great day with no altitude-related problems.*

I was acclimating well! In late August, Shawn and I took our annual Colorado 14ers trip about one month earlier than usual. The Kenya trip was on my September calendar and the dates were not flexible. In Colorado we climbed eight 14ers in a week and it was proving to be the perfect preparation for this climb to 17,000 feet. I wasn't sure how many weeks the altitude benefit from Colorado would last but the effects were still benefiting me. I think I could have run to Lenana Point and back!

Generally, we say the atmospheric air contains one-third less oxygen at altitudes above 10,000 feet than at sea level. Sometimes climbers will be tested for two oxygen-related measurements: (1) VO2 Max is a measure of how much oxygen you can take in and effectively use and (2) Anaerobic threshold is a measure of what percentage of VO2 Max you start functioning in oxygen debt.

We were not tested medically but, on this Mt. Kenya climb I felt like my body was using oxygen efficiently and never reaching the anaerobic threshold. I had experienced anaerobic states in

Mt. Kenya, 2009

some fast 5K and 10K races in the past and on Kilimanjaro's summit day. This climb feels different.

I didn't sleep much at Shipton's Camp and was wide awake at 2:00 am due to the excitement of summit day. I put on four layers of clothing—my Patagonia base layer, my moisture-wicking quarter zip cloud layer, my mid-layer down sweater, and my waterproof outer layer. I "scarfed down" some cookies and tea and left Camp at 3:00 am and summited at 5:55 am. I could have ascended the 2+ miles in two hours but we slowed down to wait for the sunrise. I summited first and got my "Climb for Sight 2009" photos.

We watched the sun lighten the skies and rise over Kenya. This was another clear, bluebird day. Two hundred miles to the south I picked out Kilimanjaro standing tall in Tanzania. The likelihood that Monday, September 28, 2009, and Monday, January 22, 2007, would both be clear mornings with 200+ miles of visibility allowing this occurrence, still blows my mind and stirs my soul! We left Mt. Kenya's Lenana Point summit at 6:30 am and arrived back at Shipton Camp at 7:45 am. *The three inches of new snow overnight had improved our footing up scree slopes.*

The weather in this region includes primarily rainfall in two seasons—March to mid-May and late October to mid-December. Rain, and higher up, snow can be encountered at any time of year, even in the driest periods of January and February. Normally the drier seasons are associated with clear, dry weather which can last for many days on end. The best weather is generally in the mornings, and convectional rainfall, if any, tends to come in the mid-afternoon.

I was initially confused by the apparent contradiction between shrinking glaciers/ice caps on high mountains and the fact that it snows up high naturally and frequently. Wouldn't those seasonal snows build up on the glaciers? Apparently, higher average temperatures melt the fresh snow.

Even when we get 3–4 inches of snow in the Texas Panhandle at 4000 feet elevation, the sun and warm temperatures melt the snow in one or two days. The sun is very intense on top of high mountains. Also, because the disappearing glaciers are so well-documented with photos and measurements, the new snow is not re-building the glaciers currently.

One related issue to melting mountain glaciers is referred to as "ice melt" and there are two types:

1. The melting of sea ice, like that of the Arctic Ocean.
2. The melting of glacial ice, by which we mean landlocked ice such as that in Antarctica and Greenland.

I have limited experience near the Poles but the same global warming trends that produce extreme weather events in the atmosphere and in the oceans affect mountain weather and glaciers. *Notice that extreme events include severe winter weather, leading to the ironic result that global warming can lead to heavy winter snowfalls.*[11]

Breakfast tasted so good—cereal, fruit, French toast, pancakes, and sausage! We packed our gear, took a few more pictures, and left Shipton Camp at 9:00 am. We hiked down to Old Moses Camp in 3.5 hours arriving there at 12:30 pm. With the summit hike and descent, we covered 12–13 miles today. Our lunch of sandwiches, noodle soup, pork n' beans, and fruit were ready for us when we walked into camp. I rested through the afternoon until MKG served popcorn, cookies, and tea at 4 pm.

I love meeting new people and the conversations we have! After telling where we are from, we usually talk about the mountains, trips we've made and want to make, and favorite places. I met a very cordial couple from Germany with plenty of experience in the Alps and European trekking.

I visited with five medical students from the University of Nairobi and told them of my association with the Lighthouse for Christ Eye Centre in Mombasa. The Clinic is always looking for medical interns and medical residents. An eye doctor named Roger DeHaven from Tyler, Texas, introduced me to the Lighthouse by calling me one day in late 2000 and asking me to meet and buy my lunch. Roger became a great friend and mentor to me.

The Lighthouse is staffed by about 40 native Kenyans who serve at all levels including the medical ophthalmologists. The clinic did not start fully funded and staffed. A California medical doctor of ophthalmology named Bill Ghrist, his wife June, and their family moved to Kenya in 1966. In 1969 they decided to move to a major city so he could focus on his specialty as an eye doctor. Nairobi, the capital city, already had several eye doctors, but Mombasa had only one aging eye doctor.

The Lighthouse site was chosen close to a General Hospital, near the Old Harbor where there is a regular sea breeze, and on a major street with great accessibility for patients. Recently celebrating its 50th year of operation, Lighthouse is known all over Kenya and the East Coast of Africa for quality eye care. When I meet Kenyans, they always seem to know the name and reputation of the Lighthouse for Christ Eye Centre.

Day Four of my Mt. Kenya climb started at 5:45 am after a good night's sleep. I packed and re-packed my gear, a climber's constant routine! Breakfast was served at 6:30 am and it was another great meal featuring a huge omelet with grilled onions and peppers, sausage, fruit, and tea. A few of us left Old Moses Camp at 7:30 am and hiked the six miles down to the Sirimon Gate by 9:00 am. Along the way we saw herds of deer (probably duiker or Cape bushbuck which are antelopes), Cape buffalo, and an elephant or two in the distance.

Big animals are not found high on the mountains but hiking through the lower forest and moorland zones climbers need to watch for them. I was surprised on our approaches to Kilimanjaro in 2007 and Mt. Kenya in 2009 that we saw so few animals. Big animals have their feeding grounds and watering holes, and they avoid humans for the most part. *How are wildlife populations being affected in east Africa?*

Climate change may significantly disrupt the ecosystem services by affecting wildlife species distribution and interspecies relationships. Kenya's wildlife species are expected to be affected in a variety of ways with changes in temperature and rainfall affecting seasonal events and species ranges. Forests cover 7.4 percent of Kenya's land and the impacts on water and water quality, wood yield and food sources, will in turn affect wildlife including wildlife habitat.[12]

Humans are already having an impact on wildlife in Kenya both indigenous populations and tourists—tourism in Kenya is a $2.5 billion sector of the economy. When we climbed Kilimanjaro in 2007 there were estimated to be 100,000 hikers and climbers annually and on Mt. Kenya in 2009 there were approximately 25,000 people annually visiting the Park.

Many animals and wildlife are confined to Africa's game parks where many tourists take a wild game safari to get a glimpse of the unique species of wildlife. Humans are impacting the earth's climate and the wildlife. One statement from the Sustainable Development Goals regarding life on land summarizes my new understanding:

Human activity has altered almost 75% of the earth's land surface, squeezing wildlife and nature into an ever-smaller corner of the planet.

As we walked out the Sirimon Gate with its World Heritage Site sign in view, I was already counting this four-day trek as one of my favorites. The World Heritage Site designation applies to over 1000 sites today like the Pyramids of Egypt or the Great Barrier Reef in Australia that are of outstanding value to humanity and to be protected for future generations. There are 24 World Heritage sites in the U.S.A., and I have visited 17 of them. I have visited another 41 World Heritage sites in 12 countries. *Just sayin' you should check it out.*

Several climbers, including our guides, shared a taxi for the thirty-minute drive back to the Ibis Hotel in Nanyuki. I had some tea and presented tips to our guides and gave them a few gifts from the U.S. I left for Nairobi around 10:30 am and arrived about 1:30 pm for a two-day stay at the Methodist Guest House (MGH) in the Lavington Gardens section of Nairobi.

I had first stayed at the MGH in 2001 when my daughter Savanah and I spent three weeks in Kenya. Savanah was 12 years old that year and wanted to join me on my first visit to a new client—Lighthouse for Christ Eye Centre! I had to promise my wife that Savanah would be safe, and we would return home!

Mt. Kenya, 2009

The first week we spent in western Kenya lodging in the town of Butere not far from the Ugandan border and the shores of Lake Victoria (and ironically, very close to Homa Bay where the Australian doctors were living). The lodge was comfortable, but it was rural and rustic in many ways. The second week we spent on the east coast of Kenya in Mombasa, lodging at the Lighthouse compound with security fences all around us.

So, I wanted our last week in Nairobi to be nice, safe, and comfortable lodging. Tim Ghrist, Mission Director of the Lighthouse, recommended the MGH. It was a beautiful place with great rooms, wonderful food in the dining hall, swimming pool, and restaurants and shopping within walking distance. We highly recommend the MGH!

As chance or divine providence would have it in 2001, I met a great man who would become a good friend for the next few years. His name was Lawi Imathiu and he was the Bishop of the Methodist Church in Kenya. Among many accomplishments, Bishop Imathiu had conceived the idea and built the Methodist Guest House. We maintained our connection through Kenya Methodist University in Meru, and we met for his board meetings in the U.S.

He was certainly a great man and a visionary, and I have so many inspiring stories from him. I once heard him speak about his heritage. He began life humbly as one of 11 or 12 children and had no shoes until he was an adult. He told the history of young missionaries from the UK coming to Kenya over 150 years before and knowing they would never return home. There were no hospitals and no roads in Kenya in those days. *Go into all the world and teach everyone you meet God's way of life!*

Bishop Imathiu described how he would have never received a formal education or had the opportunities for leadership if it had not been for those young Brits and their courage. It was a rich and meaningful talk with one main point: *We stand on the shoulders of those who came before us.*

I carry that truth in my heart and mind everywhere I go. After numerous phone calls home, flight plans changed by KLM, reading and waiting, flight delays and long flights, and one obnoxious

U.S. Customs officer, I was back in the USA! This Mt. Kenya trip was one of my best! We do stand on the shoulders of those who came before us, especially the great ones like Roger DeHaven and Lawi Imathiu in my life. We need great leaders to address climate change and reduce global warming.

4

China, 2010 and 2012

Air and Water Pollution

We never tire of looking at each other—
the mountain and I.
—Li Po

I am a numbers guy. I like to count things, measure things, calculate things, quantify things—calendar dates, mountain elevations, running miles, chronological time. Chronological is from the Greek, *Chronos*, and means "arranging events in their occurrence in time." I am not obsessed with numbers and timelines; it is just in my DNA. I am also a *Kairos* guy. To the Greeks, Kairos means the right or opportune moment. In sacred texts Kairos means *the appointed time in the purpose of God*—more like an eternal moment or a season which is qualitative not quantitative.

My first trip to the People's Republic of China occurred in *time* as January 1990 –just months after the Tiananmen Square massacre and pro-democracy protests in Beijing. My wife Lou Ann and I travelled to Hong Kong where I was attending an economic development conference and presenting a paper on *Comparative Economic Systems*. On the days when I was attending sessions, Lou Ann thoroughly enjoyed exploring Hong Kong using the clean, modern, and safe subway system. (We are disturbed by the Chinese government crackdown in Hong Kong and threats to Taiwan.)

Our economics association made it possible to enter China so soon after the events in the summer of 1989. Tensions were still high and border guards and customs officers were armed with automatic weapons. Even with prior approval and documents in hand, it took hours for our group to cross into Shenzhen/Canton, China.

We were guests of several professors and administrators of the China Education Service Center in Shenzhen. We had several dinners together and heard several presentations on education. Several tour groups took optional flights to Shanghai, Beijing, or X'ian for very short tourist visits. In 1990 the PRC was still developing a modern economy, and my initial *observations* surprised me:

1. The China we saw had many facades behind which were old buildings and poverty.
2. The clothing, culture, and environment was drab and gray.
3. Pollution in major cities was a problem.

Perhaps the contrast with Hong Kong, economically developed and modernized, was in sharp relief during our brief time there.

Returning to China twenty years later in a very different *season* or moment, we couldn't have been more amazed at the economic growth and development of the last two decades:

- Bicycles and rickshaws were now vastly outnumbered by cars and motorized vehicles.
- Thousands of new high-rise buildings have been built for apartments and offices.
- Every sector was booming, including education and transportation, which was the focus of our group.

China, 2010 and 2012

My wife Lou Ann had been selected as a Ford Foundation Fellow—one selected from each state—50 Fellows, spouses, and staff formed our group. It was planned and organized by the Ford Foundation and the American Chamber of Commerce Executives (ACCE). We met with university officials as we travelled from city to city and made tours of cultural and economic sites.

Rather than describe in detail our itinerary and agenda on this 2010 China trip, I want to weave together the two trips, the hiking and exploring from our 2012 China trip, and the picture of the climate challenges in the Chinese economy we observed.

Our daughter Mallary and her husband Chip moved to China in 2011 to teach English to university students and to study the Mandarin Chinese language. Their move, our youngest daughter's summer wedding, and professional obligations caused me to cancel my spring Denali climb as well as my annual fall Colorado trip in 2011. Lou Ann and I travelled to Changchun in the summer of 2012 for a visit and to do some hiking and exploring in Jilin province. My goal in travel is to always explore mountain terrain and the surrounding culture. We took a three-day trip to Chang Bai Shan (CBM) —-the highest mountain in Jilin.

The PRC contains many mountains, and I have in recent years been drawn to the idea of exploring them in Western China. What we usually call Tibet, the PRC refers to as the Tibet Autonomous Region with mountain names like Kailash, Kangto, and Khumbutse among many other 20,000+ feet peaks northwest of the Himalayan Range with its 8000-meter peaks.

One mountain in Xinjiang province, Muztagh Ata, has always fascinated me as well. Two books about exploring this region are both by Rick Ridgeway—*The Big Open*—a journey that he and three climbing friends took on foot across Tibet's Chang Tang and *Below Another Sky*—a journey he took with a friend's daughter to re-trace his steps and find where her father and his friend had died in the mountains of western China. China really is a vast and ancient land.

Climate change in China is having major effects on the economy, society, and the environment. China is the largest source of carbon dioxide emissions, through an energy infrastructure heavily focused on fossil fuels and coal. Also, other industries, such as the burgeoning construction industry and industrial manufacturing contribute heavily to carbon emissions.

However, due to China's large population, China's per capita carbon emissions were considerably less than the United States. In total carbon emissions, China and the U.S. are the top two countries contributing to climate change. China is suffering from negative effects of global warming in agriculture, forestry, and water resources, and is expected to continue to see increased impacts.[13]

On all three of my trips to China the obvious climate problems, like air pollution, were easy to *observe* in the cities like Beijing and Changchun with very large populations and millions of cars and trucks. Of course, air pollution can be seen in American cities like Los Angeles or Denver as well. I have made three trips to India and Delhi pollution is as bad as I've seen. People there suffer from severe respiratory health problems just breathing the dirty air!

On our 2010 China Trip we had arrived in Shanghai—a very beautiful, modern city—to pass through customs and immigration for our transfer to our connecting flight to Beijing. After a short night of rest in our Beijing hotel, we did some sightseeing at Tiananmen Square and the Forbidden City with obligatory souvenir stops. The Ford Foundation fellows and the ACCE group went back to the Marriott Hotel for briefings on education and economic topics.

The next day we drove to a Jade Factory (tourist trap stop) and then on to the Ming tombs of one of the 13 Ming dynasty emperors in beautiful hills outside the city. It was cold with snow flurries, and I loved the cold weather! After a big lunch we climbed a restored section of the Great Wall—a World Heritage Site. The Wall was 4000 miles long at one time and the 2000-year-old structure is visible without magnification from the moon!

We climbed a rather steep section of the wall where every step was covered with snow frozen solid. As we went higher the steps became more and more icy and treacherous. A few of us went on up to reach the first tower. After we got everyone safely back down that section to level ground, I left the group and climbed the Wall on the opposite mountain where the climb was less steep, and the sun was hitting the steps. I climbed up higher on the first section to reach a red pagoda tower on the wall.

> *The Great Wall of China is a series of fortifications that were built across the historical northern borders of ancient Chinese states and Imperial China as protection against various nomadic groups from the Eurasian Steppe. Several walls were built from as early as the 7th century BC, with selective stretches later joined together by Qin Shi Huang between 220-206 BC. Little of the Qin wall remains. The most well-known sections of the wall were built by the Ming dynasty (1368-1644).*[14]

By contrast with the ancient wall, we saw many modern buildings and transportation infrastructure on the 2010 trip. In Shanghai—the Urban Planning Museum, brand new universities and engineering schools, high-rise hotels, the MAGLEV trains that travel at 300mph, Pudong Airport, and many sites not associated with mountains and alpine terrain.

From 1990 (my first China trip) to 2010, China's GDP or annual economic growth rate expanded at record rates. Over the period from 1979 to 2010, the country's average annual GDP growth was 15.8%. The growth was phenomenal. So was the social cost of pollution. Consider this report by the Borgen Project: "Water pollution in China is the country's worst environmental issue":

> *Half of China's population cannot access water that is safe for human consumption and two-thirds of China's rural population relies on tainted water. Water pollution in China is such a problem that there could be "catastrophic consequences for future generations," according to the World Bank.*
>
> *China's water supply has been contaminated by the dumping of toxic human and industrial waste. Pollution-induced algae blooms cause the surface of China's lakes to turn a bright green —and greater problems may lurk beneath the surface; groundwater in 90 percent of China's cities is contaminated.*

> Water pollution in China has doubled from what the government originally predicted because the impact of agricultural waste was ignored. Farm fertilizer has largely contributed to water contamination. China's water sources contain toxic levels of arsenic, fluorine and sulfates, and pollution has been linked to China's high rates of liver, stomach and esophageal cancer.
>
> Dabo Guan, a professor at the University of East Anglia in Britain, has been studying scarcity and water pollution in China for years. He believes water pollution to be the biggest environmental issue in China, but the public may be unaware of its impact. Air pollution creates pressure from the public on the government because it is visible every day, but underground water pollution is not visible in the cities, causing it to practically be forgotten.
>
> Water pollution in China stems from the demand for cheap goods; multinational companies ignore their suppliers' environmental practices. Although China's development has lifted many out of poverty, it has also sent many others into disease.
>
> Factories are able to freely discharge their wastewater into lakes and rivers due to poor environmental regulations, weak enforcement and local corruption. Rural villages located near factory complexes rely on the contaminated water for drinking, washing and cooking. These villages have become known as "cancer villages" because of high rates of cancer and death.[15]

In the summer of 2012, we took our third trip to China, and I insisted that we get in some hiking and climbing while spending time with family. Chip and Mallary were great hosts providing us with real cultural experiences in northeast China. We met their Chinese students, ate in small restaurants where they knew the owners and ordered local favorites like Guang Do chicken or the Three Treasures dish with potatoes, onions, and peppers, and explored Changchun. When will we go to the mountains?

Finally, we got bus tickets and took a six-hour bus ride east from Changchun toward the Changbai and Baekdudaegan mountain ranges in northeast China and North Korea. We lodged the

China, 2010 and 2012

first night of three nights at the Woodlands Youth Hostel in Erdaobaihe. The mountain range straddles the border with North Korea and the highest point cannot be hiked on the Chinese side of the border. On the second day there (7/4/2012), we took more buses and vans to the trailhead leading up Changbai Mountain.

The North Koreans know it as Paektu (Baektu) Mountain, and it is an active stratovolcano. Koreans assign a mythical quality to the volcano and its caldera lake, considering it to be their country's spiritual home. At 9003 feet (2744 meters), it is the highest mountain in Korea and Northeast China. It was regarded as the most sacred mountain in the shamanist religion and many people have done some hiking there. There is no record of the first ascent (FA) of Mount Changbai.

We hiked up above 2700 meters in rain, fog, and high winds. North Korean soldiers guard the invisible border. Mallary and I hiked the section of the summit loop on the Chinese side twice but never caught a glimpse of Heaven Lake in the summit caldera. Evidently, fog and clouds frequently hide the lake.

You are not considered unlucky or cursed if you cannot see it, but you are considered a very lucky person if you get a view of the lake, and good things will follow in your life. The Changbai Shan Park is a Chinese tourist destination and a very "controlled" environment partly due to the mountain border with North Korea.

The weather is typically filled with clouds and fog, partially due to rising levels of magma below the central part of the mountain. The volcano last erupted in 1903, and scientists expect it to erupt about every 100 years. The weather on the mountain can be very erratic, sometimes severe. The average temperature remains below freezing for eight months of the year. The lowest record temperature was -60 degrees Fahrenheit on January 2, 1997.

Returning to the trailhead, we left the road and motorized transportation and hiked four miles to the spectacular Changbai waterfalls. There were fewer tourists here and it reminded me somewhat of Yellowstone NP with its steam vents, waterfalls, and landscape. Chip and Mallary told us to bring swimwear but did not tell us why.

They led us to the Julong Hot Springs with a Men's Bathhouse and a Women's Bathhouse for a soothing sit-down in the natural springs. The local tradition is one of nude bathing in the hot springs and I respected the custom with only four or five Chinese men in the Men's bathhouse.

Lou Ann and Mallary kept on their swimsuits while sitting with 25–30 women on the women's side. We laughed about this surprise experience many times, but the funniest story was on the women's side. The Chinese women were all local schoolteachers who were, of course, speaking local Mandarin and calling the two American women prudish and modest for not bathing nude. They didn't know that Mallary understood most of what they were saying. Before they left my daughter struck up a conversation with the women in their language. Talk about surprise! I like the old saying, "There are no foreign places, only the traveler is foreign."

It was good to be in the mountains of northeast China. The air was crisp and clean to breathe, and the population was much lower. The sheer numbers of people in China's cities creates pollution and environmental impacts. Today China's population exceeds 1.4 billion people and China has ten cities with over eight million people. China has 50 cities over 2 million and Shanghai alone has almost 27 million people. The human impact spread out over a large country is still significant and *observable.*

> *Efforts to curb air pollution in China, a country already facing dire health impacts from high levels of soot and smog, will likely become increasingly difficult as the planet warms according to the new studies. Increased heat waves and more periods of stagnant air resulting from global warming will worsen existing air pollution across much of China scientists conclude. This presents a heightened challenge for a country already choking on airborne pollutants that cause more than 1 million premature deaths there each year.*
>
> *For Chinese policy makers working to improve current air quality and protect public health, this finding is a daunting conclusion. It is one that underscores the need to tackle*

the challenges of both climate change mitigation and air quality at the same time.[16]

During the five years that Chip and Mallary lived in China between 2011 and 2016, they lived in three different cities of very different sizes—Qufu, Changchun, and Siping. Qufu, the birthplace of Confucius in Shandong province, is a smaller city listed today as having 188,000 people in its urban area. Changchun, the capital city of Jilin province, had a population of 7.7 million in 2010.

Siping, a prefecture-level city in Jilin province, had a 2015 population of 3.3 million with 680,000 people in its urban area. Their first child, Pilgrim, was born in Changchun. After their two-year Chinese language school, they were open to moving or being assigned to another university in another city. The city of Beijing had the most air pollution and, despite short visits there, they did not want to live there and were assigned to Siping.

> *The Paris Climate Agreement is an agreement within the United Nations Framework Convention on Climate Change (UNFCCC) dealing with greenhouse gas emissions mitigation, adaptation, and finance starting in the year 2020. The Agreement aims to respond to the global climate change threat by keeping a global temperature rise this century will below 2 degrees Celsius above the pre-industrial levels and pursue efforts to limit the temperature increase even further.*
>
> *As of January 2021, 194 states and the European Union have signed the Agreement. 189 states and the EU, representing about 97% of global greenhouse emissions, have ratified or acceded to the Agreement, including China and the United States who together account for about 40% of emissions. China (20.09%) and the U.S. (17.89%) are the countries with the 1st and 2nd largest carbon dioxide emissions.*[17]

China's industrialization and rapid economic growth over just a few decades had produced enormous pollution and high social costs including the externalities referenced. A few days after our return in 2012 from our 3rd China trip, Mallary emailed us four questions to answer about our time in China for her blog.

1. What most surprised you on this trip or past trips about China? *I responded that it was China's explosive economic growth.*
2. What do you most like about China? *I liked meeting the Chinese people and sensed in many of them a hopeful spirit.*
3. How is China different from the USA? *Eastern culture and worldview differ from Western culture. China has a longer history and traditions than the USA.*
4. What was the most difficult aspect of this trip to China? *In the cities I found the huge number of people and crowded spaces difficult. I would add today that the air and water pollution is difficult.*

The observations about air and water pollution made worse by climate change and global warming in China demonstrate the interconnectedness of the work needed to reduce greenhouse gases and carbon emissions worldwide. The effort required must be a combined effort involving all polluters and countries, not just one or two. Please contact me if you are planning a hike or climb in western China or the Tibetan Plateau and need another team member. *Everything in its right time! There is a season!*

5

Carrauntouhil, Ireland, 2012

Rising Temperatures

Two people make a shorter road.
—Sign at the Cronin's Yard TH

When you have four grown children and now some grandchildren living in four different places, you travel to see them. Frequently that travel has been international because all of them have lived or spent extended time out of the United States. Before our China trip in the summer of 2012, we flew from Chicago to Dublin in April 2012 to celebrate Easter and spend a week touring that country.

Our daughter Savanah and her husband Scott were living in Galway and drove across Ireland to meet us. We spent a day exploring Dublin including the Guinness St. James's Gate Brewery and then drove across the Emerald Isle to Galway. We attended Easter services in the Galway Cathedral, a Renaissance Revival style of architecture, and with only a few people in attendance.

We loved the quaint county town of Galway with its Eyre Square and numerous venues and events. We lingered at the farmers market, tasting the cheeses and jams. After two lovely days in the charming west coast town of Galway walking and running to stay in shape, we began our Irish driving and hiking adventure.

Our plan was to drive around the Emerald Isle sightseeing, and my mountain plan included climbed the highest mountain in Ireland—Carrauntoohil. We decided that Scott and I would share the driving which is on the opposite of the road from our U.S. roads. You adjusted to it quite quickly, but I was happy to enjoy the lush green scenery as a passenger more than I could as the driver. Our first stop included the Cliffs of Moher and ancient ruins of castles, abbeys, and churches on the Dingle Peninsula.

We stayed the first night at the Dingle Gate Hostel in Anascaul not far from the village of Dingle. Anascaul and the County Kerry were the home of Tom Crean, Antarctic explorer and survivor. I found a copy of Michael Smith's book *An Unsung Hero* and began to read about this guy I had never known. Sir Edmund Hillary said, "Tom Crean was a great man of immense strength and endurance and afraid of very little." Irish rugby player Trevor Brennan said of *An Unsung Hero*, "One of the most inspiring books I've ever read." From the preface of *An Unsung Hero*—

> *The Dingle Peninsula, Kerry, is one of Ireland's most beautiful spots, its rich mixture of rolling hills and rugged coastline jutting out into the Atlantic, nature at its best. Visitors today come from all over the world to admire the dramatic scenery. About midway along the Peninsula, in a modest uncomplicated setting, sits the small village of Anascaul. It is said that Ireland's last wolf was killed in the overlooking hills. But visitors passing through the main street of Anascaul are alerted by one of the last buildings they glimpse as they travel west towards the better-known town of Dingle and the Atlantic breakers.*
>
> *This is a small pub with a highly unusual name: the South Pole Inn—situated alongside a quietly flowing river and a charming stone bridge, nowhere on earth seems farther from the South Pole. But it is impossible to arrive or leave Anascaul without catching sight of the little building and wonder how a public house in a rural village surrounded*

Carrauntouhil, Ireland, 2012

by endless green fields on the Dingle Peninsula came to be called the 'South Pole Inn'.

The answer, for those who linger, can be found in a small slate-grey plaque above the pub doorway. It reads:

Tom Crean Antarctic Explorer 1877-1938

The South Pole Inn was the home of Thomas Crean, a local man who rose from the obscurity of a typical farming community in Kerry to become one of the greatest characters in the history of polar exploration at the turn of the century—the Heroic Age of polar exploration.[18]

Rising early on the first Tuesday after Easter Sunday, Scott, Savanah and I climbed the hills above Anascaul Lake. We had an awesome bluebird morning with great weather, spectacular scenery, and landscapes made for photographing. Long after I have lost or misplaced the photos, this place will remain imprinted in my memory.

Collecting Lou Ann from the Hostel where she had enjoyed a warm fire all morning, we continued our driving tour. *My wife has Cherokee blood, and their society had a central philosophy, called 'duyuktv', which means "the right way." It focused on seeking balance and harmony in life and respecting the natural world. We see it embedded in their central plazas and council houses—home to the sacred fire.* On this particularly cold April morning, Lou Ann chose the warm fire.

The Geological Survey summarizes the effects of climate change in Ireland:

> *The effects of climate change can be clearly seen, the most evident effect can be seen in the increased temperature between 1890 and 2008 and more significantly between 1980 and 2008. These increased temperatures have had knock on effects on Ireland's natural environment; it has changed the growing season affecting farming and has increased the number of animals suited to warmer temperatures. An increase in the frequency and impact of storms has also been recorded in the last few decades.*

MOVING MOUNTAINS

> *If the rate of global warming continues to increase and the climate continues to change, there could be adverse effects in Ireland. As an island nation we are particularly vulnerable to increasing sea levels with coastal regions facing issues of flooding. More erratic weather conditions could lead to both water shortages as well as impacting water quality. Changing weather could also have devastating effects on the plants and wildlife of the country.*[19]

We lingered all afternoon and saw even more amazing emerald landscpes and breathtaking views on the Ring of Kerry. We stopped in the town of Killarney for lunch and then walked several miles in the town. Due to its proximity to Carrauntoohil, we booked a room at Darby O'Gill's Country House Hotel and Pub in Killarney.

The next morning Scott, Savanah, and I drove to Killarney National Park and the Macgillycuddy's Reeks Mountain range to climb the 3406-foot Carrauntoohil Mountain. As Ireland's highest mountain, it is popular with mountain walkers, who most commonly ascend via the Devil's Ladder route; however, it is also climbed as part of longer mountain routes like the Coomloughra Horseshoe route and the Reeks Range Walk of the entire mountain range.

Carrauntoohil has a deep corrie, known as the Eagle's Nest, at its northeast face, which is accessed from the Hag's Glen. Sometimes the term Eagle's Nest is used to refer to the small stone Mountain Rescue hut that sits on the first level of the corrie, where the Heavenly Gates descent gully meets the Eagle's Nest corrie. Carrauntoohil is regarded by the Scottish Mountaineering Club as one of 34 Furths (mountains above 3000 feet) and is one of the thirteen Irish Munros (a Furth outside of Scotland).

We parked near the trailhead at Cronin's Yard and started on the easy approach to the Devil's Ladder Trail. The roundtrip distance is listed as 7.1 miles, and it is an up and back route. The trail leads straight toward the mountain, crossing a small river, passing between Lake Gouragh and Lake Callee. At the base of the steep gully called the Devil's Ladder, the climbing gets tougher on loose

Carrauntouhil, Ireland, 2012

scree slopes. On the April day when we climbed, it was raining steadily as we reached and started up the gully.

It reminded me of some higher mountains like Borah Peak in Idaho where for every two steps up you slide back down one step. Added to the steep slope were slippery, wet conditions that made you work hard and watch every step. At the top of the Devil's Ladder, you reach the summit ridge, curving right and proceeding up the ridge to the top where there is a large iron cross marking the summit.

In the 1950s, a wooden cross was erected on the summit by the local community. In 1976, the wooden cross was replaced by a steel cross. In 2014, the cross was cut down by unknown persons in protest against the Catholic Church, but it was quickly re-erected. *Some live as enemies of the cross.*

The winds were blowing strong, and it was snowing as we ascended the ridge and reached the summit. We took a few quick pictures and started back down. Carrauntoohil also reminded me of Mt. Hood in Oregon as both are small but potentially vicious mountains with their own nasty weather. This is, evidently, the typical weather pattern in this mountain range, taking note of features with names like Howling Ridge (not a hill-walkers descent route). The first ascent (FA) is not recorded.

My journal entry indicates that we covered eight miles roundtrip between 7:00 am and 12noon with some additional walking around at the end. I was particularly proud of Savanah for climbing this wet, windy, and snowy peak! Scott and I had enjoyed hikes and climbs in the British Isles so much that we planned to return in 2017 to climb the high points of Wales, England, and Scotland. *Here's an excerpt from my first mountain book entitled Retreat Upward with the full story to follow in Chapter 9:*

"Scott and I have formed a great son-in-law and father-in-law relationship through hiking and climbing. Our Carrauntoohil climb in 2012 on a windy and snowy day created a mutual desire to do the Three Peaks Challenge in Wales, England, and Scotland. During a rainy week in September 2017, we successfully completed Mt. Snowdon, Scafell Pike, and Ben Nevis together."

As we continued our drive around the southern loop of Ireland in 2012, we saw so many famous and fabulous sites—Muckross House and Gardens, Blarney Castle at Cork, and Waterford town with its Waterford crystal. We visited the Glendalough monastery in Wicklow, and many beautiful landscapes and seascapes. Carrauntoohil was the highlight for me but that took only one day, and the other days were filled dawn to dusk. Two people may make a shorter road, and the four of us were great travel companions.

I was inspired by the Trinity College Library containing The Book of Kells, sometimes known as the Book of Columba. It is an illuminated manuscript and Celtic Gospel book in Latin, containing the four Gospels of the New Testament. The book of Kells was created in a Columban monastery in either Ireland or Scotland around 800 AD.

The text of the Gospels is largely drawn from the Vulgate as well as several passages drawn from earlier versions of the Bible known as the Vetus Latina. The Kells is regarded as a masterwork of Western calligraphy and the pinnacle of Insular Illumination. The manuscript takes its name from the Abbey of Kells, County Meath, which was its home for centuries.

I was also inspired by the story of the Guinness family, told in detail in the book by Stephen Mansfield entitled *The Search for God and Guinness: A Biography of the Beer that Changed the World.* Mansfield stirs together several strains of the Guinness legacy —- beer, family, philanthropy, and vibrant faith devoted to lifting the helpless and downtrodden. In an age of corporate irresponsibility and corruption, the Guinness story is a challenge to our times and an inspiration for our lives.

The fact is Arthur Guiness, founder, was indeed a great man of faith. Born on the estate of an archbishop and raised a loyal son of the Irish church, Arthur lived by the words that were his family motto: *Spes mea in deo est (My hope is in God)*. He was influenced by the revivalist John Wesley, who inspired him to use his wealth and talents to care for the hurting people. Taking Scripture as his guide, Arthur did indeed serve the needy of his time and did indeed use his gifts in honor of God."

Carrauntouhil, Ireland, 2012

The Earth is the Lord's and so too its fullness. Weather and climate are always changing. My hope is that global warming and extreme climate change and the corporate greed that causes so much damage to the Earth can be reversed. The Emerald Isle must be kept brilliant for future generations. It will require great leadership and our *cooperation.*

To quote G.K. Chesterton: *The tourist sees what he has come to see, the traveler sees what he sees.* And of course, Oliver Wendell Holmes: *The mind that is stretched by a new experience can never go back to its old dimension.*

6

Colorado 14ers, 2013

Violent Weather

---

*It is good to have an end to journey toward;
but it is the journey that matters in the end.
—Ernest Hemingway*

In September 2012, Shawn, Scott, and I did three Colorado 14ers—Mount of the Holy Cross, Mt. Elbert in a blizzard, an attempt on La Plata Peak. We backpacked over Halfmoon Pass and camped near East Cross Creek. We climbed the north ridge route to the top of Holy Cross and the second day and hiked back out to the Halfmoon trailhead that day. I will tell the Elbert and La Plata stories later.

Late August and early September 2013 came around quickly, so Shawn and I decided to climb some of the tougher Colorado peaks in the Elk Range. I knew I would need an experienced climber to join me. We finished two peaks and then the monsoon rains began and chased us east to the Sawatch Range.

Moving Mountains

Weather is always a factor in the Rocky Mountains and, at times, an unpredictable event. Mountains can create their own weather. Counting several neighboring peaks, we would stand on top of nine peaks over 14,000 feet high during this week-long trip. What is a Peak?

Gerry Roach, author and climber, answers that question in his guidebook *Colorado's Fourteeners: From Hikes to Climbs*:

> The traditional list of Colorado fourteeners varied from 52–55 peaks (Some folks use a list of 58) over the years and has always been based on a healthy degree of emotionalism. Peaks have come and gone for sentimental reasons. In this high-tech peak-bagging age, however, many climbers are interested in peak lists based on a rational system. These climbers are interested in the discussion about what constitutes a peak because the answer determines a list's contents and their climbs.
>
> For some time in Colorado, a single, simple criterion has been used to determine if a summit is a peak or a false summit: For a summit to be a peak, it must rise at least 300 vertical feet above the saddle connecting it to its neighbor peak. If just one criterion is to be used, this is a good one. There is nothing sacred about 300 feet. It is just a round number that seems to make sense in Colorado.[20]

Shawn has lived in northern Virginia for the past 10–11 years and I have lived in Arkansas during that same time, so we usually book our flights through the Denver International Airport, meet there, rent a car, and make stops at an REI store and grocery market before driving to the trailhead (TH) camp site. I am always eager to get out of the polluted and congested Denver traffic and into the nearby mountains.

In 2013 our TH destination was Castle Creek TH for access to both Castle Peak and Conundrum Peak. These peaks are 12 miles south of Aspen and are usually climbed together. The Castle Creek TH is located at 9800 feet and there are some campsites there. We arrive late afternoon or early evening from our respective homes at sea level.

Shawn and I both exercise year-round to stay fit for the mountains and we tend to adjust well to altitude. The first day of climbing is always the hardest. We have never needed a full or extra day to acclimate but as I pile up more years, I'm thinking that strategy would be wise.

We hit the trail early the next morning leaving the TH sign at 5:00 am on the Northeast Ridge route which is a ten miles roundtrip hike with 4500 feet of vertical gain to the Peak's summit at 14,265 feet. Rather than bore the reader with every route description (which you can find online or in a guidebook), I will instead give a general explanation of our experiences with these two peaks. This Castle Peak route has good trails to the top, is rated an easy climb after the snow is melted by mid-summer, and only requires some Class 2+ scrambling. Here is a typical mountain climbing rating:

- **Class 1**—trail hiking with 3000 feet vertical gain.
- **Class 2**—off-trail hiking with vertical gain on a talus slope.
- **Class 3**—the easiest climbing category called "scrambling." You use handholds and basic climbing techniques.
- **Class 4**—the territory of technical climbing: handholds and foot placement that must be tested. Use upper body strength.
- **Class 5**—is technical climbing. You use climbing techniques and sometimes technical equipment—harness, helmet, ropes.

Shawn and I summited Castle Peak at 10:00 am on 8/29/13. We crossed the connecting saddle ridge descending and then up to summit 14,060-foot Conundrum Peak and descended to the Castle Creek TH at 4:00 pm. We summited two 14ers on Day One and did it easily to acclimate to altitude.

Since our plans were to climb the seven peaks in the Elk Range on this trip, we headed for the Maroon Lake TH with access to Maroon Peak, North Maroon Peak, and Pyramid Peak ten miles west of Aspen. The National Forest camping sites were apparently full but approaching dark rain clouds were even more threatening as we saw the famous postcard view of the Maroon Bells.

Checking the weather forecast, this event was not going to be a summer thunderstorm but a multi-day monsoon rain. *The rain falls on the just and the unjust.* We decided to go back east over Independence Pass and across the mountains to the Mt. Massive area. Hopefully, we will hike there before the rain advances. I discovered this climate article in early 2021:

> *Aspen mayor Torre (he goes by one name) says "our snow is running out. That's a major problem when you're a ski resort." The numbers are, frankly, alarming. Aspen already gets a month less of skiing than it did in the 1940s, with snow levels across Colorado having receded by 20 to 60%. If drastic action isn't taken, the EPA warns that the standard ski season will be halved by 2050.*
>
> *"What makes Aspen Snowmass stand out is that they were one of the first major resorts to be vocal about seeking climate solutions." Many of the U.S.'s major mountains—including Steamboat and Deer Valley—have joined resorts in Aspen in lobbying for policy change in Washington, DC.*[21]

We arrived at the Mt. Massive TH by evening and set up our tent. Mount Massive is 11 miles southwest of Leadville and the TH is at 10,050 feet. Mount Massive is the 2nd highest Colorado 14er just 12 feet shorter than Mt. Elbert right next door. Massive has five summits above 14,000 feet on its three-mile-long summit ridge and no other mountain peak in the lower 48 states has a greater area above 14,000 feet.

The next morning (8/30/13) we left the TH at 6:00 am for our 13.6-mile roundtrip with 4450 feet of vertical gain. We summited Mt. Massive at 10:45 am and then went back across the connecting saddle to South Massive and down to TH at 3:30 pm. We summited two more 14ers today and it was a "win" that erased two previous "losses" on Mt. Massive. This is the back story:

Following a conference at the Broadmoor Hotel in Colorado Springs in October 2008, two friends and I drove up to the Leadville area to climb Mt. Massive. We got a late start, and my climbing companions hiked slower than I did. Mid-afternoon I was near the South Summit when we hit our turnaround time, and I went back down to join them, and we hiked back out.

In September 2010 I flew to Denver and drove straight to Half Moon Creek TH for a quick climb of Mt. Elbert. I summited but I did not acclimate at all, and Elbert took every ounce of strength I had. It was my slowest roundtrip time on Elbert in four ascents over the years including my 1999 climb with my 81-year-old Dad. The very next day I started up Mt. Massive and after about four miles realized I was done. Mt. Massive—2; Tom—0. In 2013 I summited Mt. Massive and South Massive!

Never Stop Exploring and Never Quit!

We drove south toward Buena Vista and then west to the West Winfield TH at 10,380 feet with access to La Plata Peak's south side. La Plata means "silver" in Spanish and is Colorado's fifth highest peak. It is about halfway between Twin Lakes and Independence Pass south of Colorado 82.

We made camp in one of the informal campsites and heated water for a freeze-dried dinner. This wasn't our first attempt at La Plata. In 2012 we had started up the trail, missed a turn due primarily because we were not paying attention, and ended up climbing an unnamed peak east of La Plata above 13,000 feet. We would correct that mistake this year.

Shawn and I left the TH at 7:00 am on 8/31/13 and took the Southwest Ridge route, a gentle, scenic Class 2 trail to the summit. We each hiked at our own pace and Shawn reached the top first at 10:30 am with me topping out at 10:45 am. We spent a short time on the summit and *observed* clouds building up and moving toward us. It was time to get off the mountain!

I would climb Culebra Peak in southern Colorado in 2016. On that climb I met another climber, Henry, from Castle Rock, CO, who was completing his 55th Colorado 14er. We hiked together and talked a lot about climbing and summer thunderstorms with lightning in the Rocky Mountains.

I shared my closest call with lightning which occurred on La Plata in 2013, and he told me his closest call had been a few years before on Capital Peak. Lightning is the number one cause of death in the high country and is notorious in July and August.

Experienced climbers try to get an early start and get off the summit and down below the treeline by midday. I've had one good friend, Billy McCauley, killed by blue sky lightning in the Pecos Wilderness of northern New Mexico. This is the characteristic Rocky Mountain weather to which climbers and hikers must pay attention.

Before Shawn and I could descend the summit ridge to the Southwest ridge, we were hit by violent storms first pelting us with rain and hailstones and then snow flurries. We got separated from each other. Shawn found shelter beneath the overhang of a large boulder. I just continued to descend as fast as I could to look for shelter when I reached the cover of the trees. Thunder was crashing and lightning was electrifying the air around us.

These kinds of thunderstorms in the Rocky Mountains are common and not new. I have had close calls on other mountains as well in Arizona and California. One or two experiences of violent weather were not proof of global warming. What I want to be very aware of as a climber and hiker are unexpected heavy downpours and extreme weather events increasingly triggered by climate change. The drought and wildfires in California during the summer of 2020 and the polar temperatures and heavy snowfall in Texas during February 2021 were both extreme weather events.

After the storm passed, we looked for each other further down the mountain. I waited for Shawn to descend as I assumed he was above me on the mountain. We would ask people if they had seen a climber dressed as we described. Finally, we re-connected down the valley where several trails converged. We knew we should always stay together. However, this crazy storm blew up so fast. We returned to the TH at 2:30 pm and rested in our tent. We moved camp to the South Winfield TH at 10,260 feet providing access to the north and west sides of Huron Peak.

Huron Peak has been described as a shy peak with a pyramid-shaped summit that just barely rises above 14,000 feet. It is the farthest Sawatch Range fourteener from a paved road and yet, it is an easy Class 1 climb with 3740 feet of gain on a roundtrip trail measuring eight miles. We left the TH on 9/1/13 at 5:30 am taking the northwest-facing slope to the summit arriving separately—Shawn at 7:55 am and Tom at 8:15 am.

The views from Huron's summit are maybe the best view in the Sawatch. Following a snack and pictures, we descended to camp arriving at 10:30 am. We felt fully acclimated at this point and were ready to tackle not a "two-fer" but a "three-fer" the next day. [1]

For that challenge we drove to the Missouri Gulch TH at 9640 feet eight miles west of U.S. 24 between Buena Vista and Leadville. The three mountains here are Mt. Belford (14,197), Mt. Oxford (14,153), and Missouri Mountain (14,067) and they can be hiked in one long day as a combination climb. If we kept moving at a steady pace, this would be a twelve-hour hike.

We left the TH on 9/2/13 at 5:00 am and climbed Missouri Mountain's Northwest Ridge Route reaching Missouri's summit first at 9:00 am. We then descended a short distance and traversed to the east to summit Mt. Belford at 11:45 am. We had a light lunch and water. Leaving at noon, we continued east to Mt. Oxford, summiting at 12:50 pm. After a brief rest, we hiked west back over Mt. Belford and down to camp by 4:40 pm. We had summited nine peaks in five days, and we were "on a roll."

With another day to climb before returning to Denver, we re-packed gear and drove west of Buena Vista to the North Cottonwood TH at 9880 feet for the approach to Mt. Harvard. We found a great campsite at North Cottonwood and enjoyed a relaxing evening in the glow of our alpine achievements.

On the morning of 9/3/13 we were anticipating 14er #10 on this trip. It was not to be. A few miles up the NC trail one of the muscles in my right leg locked down. Was it an overuse injury like I had experienced in competitive running? I made the decision to turn around. Shawn agreed.

We'll get it next year! Climb with humility and respect. The mountain will be there.

This trip was awesome! We managed to bag two peaks in the Elk Range before the monsoon rains arrived. We moved east over the mountains, faced my unfinished business on Massive, and added two more peaks. We were able to get La Plata and survive the sudden storm and then do Huron after the storm had moved on. We finished the Missouri-Belford-Oxford combo making a total of nine 14ers.

Ending our Mt. Harvard attempt early meant we didn't have to hurry back to Denver to catch our flights. We could linger in the mountains, go back down to civilization in Buena Vista, and eat at one of my favorite mountain town breakfast places. Here is the short list:

1. Evergreen Café, Buena Vista, CO
2. Michael's Kitchen, Taos, NM
3. Schat's Bakery, Bishop, CA
4. Sierra Grande Café, Des Moines, NM
5. Sylvan Lake Lodge, Custer, SD.

Mt. Harvard is the third highest summit in the Rocky Mountains of North America and the state of Colorado, named in honor of Harvard University. The first ascent (FA) recorded was on August 19, 1869, by members of the first Harvard Mining School while on expedition. Several mountain names include prominent universities such as Yale, Princeton, Columbia, and Oxford, all in the Sawatch Range.

The Collegiate Peaks or Range is the name given to a section of the Sawatch Range of the Rockies located in central Colorado. Drainages to the east include the headwaters of the Arkansas River. Since moving to Arkansas and living near the Arkansas River almost twenty years ago, I have become very interested in in this great American river. My wife served as the President of the Arkansas Waterways Association for several years and we travelled from its source in Colorado to its convergence with the Mississippi.

The Arkansas River is a major tributary of the Mississippi River and generally flows to the east and southeast as it traverses the states of Colorado, Kansas, Oklahoma, and Arkansas. The river's source basin lies east of Leadville, Colorado, and creates the

Arkansas River Valley flowing south. The headwaters derive from the snowpack in the Sawatch and Mosquito Mountain ranges. In the Arapaho language the Collegiate Peaks were called the Elk's Head.

In earlier times the Collegiate Peaks Wilderness area was inhabited by various people groups. The native Utes used the bark from the plentiful Ponderosa Pine trees for clothing and food. This area is also dotted with the evidence of mining operations from the 1800s. In the Pine Creek valley—one of the central valleys between Mounts Oxford, Belford, and Missouri to the north and Mount Harvard and Columbia to the south, there is evidence of an early settlement comprising cabins and a corral as well as other ghost towns.

Varied factors bring changes to the mountains: economic forces, population shifts, as well as weather and climate. The possible climate change effects mentioned in this chapter including extreme weather events, heavy downpours, and reduced snowfall are not the only effects occurring in Colorado. I will summarize the others in Chapter 11 on another Colorado 14ers trip to the Chicago Basin in the San Juans. This 2013 trip was truly a "Rocky Mountain High" (John Denver)—*Talk to God and listen to the casual reply!*

7

Coma Pedrosa, Andorra, 2014

Water Resources and Snow Cover

Wherever you go, go with all your heart.
—Confucius

Andorra is the sixth smallest nation in Europe with an area of 181 square miles and a population of about 85,000. It is the 16th smallest country in the world by land and the 11th smallest by population. The capital is Andorra la Vella and it is the highest capital city in Europe at an elevation of 3356 feet. Tourism is the largest sector of Andorra's economy. In 2013, Andorra had the highest life expectancy in the world at 81 years. Located in the eastern Pyrenees, Andorra is bordered by France to the north and Spain to the south.

Due to its location in the Pyrenees Mountain range, Andorra consists predominantly of rugged mountains, the highest being Coma Pedrosa at 9656 feet and the average elevation of the country is 6549 feet.

Moving Mountains

Andorra has alpine, continental, and oceanic climates, depending on altitude. Its higher elevation means there is, on average, more snow in winter and slightly cooler than average summer temperatures. The climate of the Andorran Pyrenees is dominated by the high mountain climate and influenced by the Mediterranean climate.[22]

Our trip to Europe in May 2014, particularly Spain and Andorra, was characterized by excellent weather with very little rain. I have never climbed in the Alps but the climb in the Pyrenees gets me closer! Our daughter Savanah and her husband Scott were living in Barcelona and working there. My wife and I just had to visit and so we pitched the idea of a weekend trip to climb Coma Pedrosa and see the country of Andorra.

Our Delta flight was a direct one from Atlanta to Barcelona arriving with plenty of time to explore Barcelona on foot including La Sagrada Familia, Carmel Hill, and Parc Guell. We had made a few weekend excursions to Barcelona in 1997 when our family was spending the summer down the coast at Altea (the Costa del Sol).

I was the college dean of a summer session there with 55 U.S. students studying Spanish, art, international economics, and taking other courses. The University was housed in a beautiful villa right on the Mediterranean coastline. Barcelona is an amazing and vibrant city with so much to see and explore.

In 2014, we arrived on a Thursday and so the very next day we loaded the rental car and headed for the Pyrenees mountains. The drive was mostly north toward Andorra and the scenery was spectacular! We had to take a short side trip to Montserrat mountain range in Spain and it was an unexpected highlight of the weekend journey. Montserrat is a multi-peaked range near Barcelona in Catalonia and is a part of the Catalan Pre-Coastal Range. The main peaks are Montgros, Miranda de les Agulles, and Sant Jeroni with the highest point being 4055 feet.

It is the well-known site of the Benedictine abbey, Santa Maria de Montserrat, which hosts the Virgin of Montserrat sanctuary which is breathtaking. The monastery was founded in the 11th century and rebuilt between the 19th and 20th centuries and rests

Coma Pedrosa, Andorra, 2014

on the side of the mountain. "Montserrat" literally means "saw (serrated, like a handsaw) mountain" in the Catalan language. That describes its peculiar features with multitude rock formations and spires which are visible from a great distance.

The mountain is composed of strikingly pink conglomerate, a form of sedimentary rock. The summit is accessible by hiking trails from the monastery and is part of the GR footpath 172. The Cavall Bernat rising to 3645 feet is an important rock feature for rock climbers. An interesting historical fact I learned is that in 1493, Christopher Columbus named the Caribbean Island of Montserrat after the Virgin of Montserrat and this same monastery. *I wish we had planned more time in Montserrat, Spain!*

We arrived in Andorra after crossing the border and showing our passports and drove on into the quaint and captivating capital city of Andorra la Vella. We checked into the Anyos Park Mountain and Wellness Resort in La Massana, one of the seven parishes of the Principality of Andorra, and this would be our lodging for three nights. The resort had a great gym, pool, and spa and was within walking distance of town. We still had time to walk two or three miles and explore this charming place.

Andorra is a sovereign landlocked country on the Iberian Peninsula, in the eastern Pyrenees, bordered by France to the north and Spain to the south. Believed to have been formed by Charlemagne (Charles the Great), two co-princes currently head it: the bishop of Urgell in Catalonia and the president of France. Catalan is the official language in this small country of only 85,000 people.

Skiing, hiking, cross-country running, and cycling are all popular sports tourism activities in Andorra. Andorra's economy is reliant heavily upon tourism. The ski season traditionally runs from late November to early April, depending on weather conditions. Outside of the ski season, some of the ski lift facilities continue to operate to take tourists to spectacular views from mountain peaks like Peyreguils, Tristaina, Creussans, and Cabanyo.

Early on the first morning after our arrival in Andorra, Scott and I headed for the trailhead at Arinsal, another one of the seven parishes of Andorra and started up the mountain. I remember it was a little cold at first, but it warmed up as the sun got higher in the sky on a bluebird day, perfect weather for climbing an unfamiliar mountain.

Scott climbed with me while there was no snow on the lower trails, and we reached a closed ski lodge structure after a couple of hours. He turned back there because he had worn the wrong shoes for snow-covered trails. Scott would love this mountain in winter because he loves snow skiing. He reminds me of my 20-something self!

I wore my dependable Asolo alpine boots and continued in the snow all the way to the top of this snowy mountain.

> *In 1946 the Zanatta family started making boots in a little shop in Nervesa della Battaglia, Italy. The family business grew after the Second World War and really flourished in the 1960s and 1970s as the second generation of the Zanatta family continued the values of their parents. The Asolo brand was founded in 1975 by Giancarlo Tanzi who invented the first trekking boots using Cordura materials and he launched Gore-Tex lined footwear. The third generation Zanatta family bought the Asolo brand in 1998.*[23]

Thanks, Asolo and Zanatta family for making great boots! I reached the summit of the 9656-foot snow-covered Coma Pedrosa before noon and went back down by 2:00 pm with a roundtrip time of seven hours. With such beautiful weather and spectacular scenery in this very small country, I wondered if the UNFCCC had recorded any potential effects of climate change for Andorra. Here is what I found on the World Bank Climate Change Portal:

> *Andorra is a mountainous country, and therefore it is particularly sensitive to climate change. The climate of Andorra is a humid mountain climate of mid-latitudes, but with a Mediterranean influence in the southern sector, where the characteristics are of continental Mediterranean climate. Some effects of climate change are already being perceived in the country's mountains with an increase of about 0.17°C per decade in temperature and a decrease in annual rainfall of about 49 mm per decade. These*

Coma Pedrosa, Andorra, 2014

variations are likely to result in impacts on water resources and snow cover, essential for sports related tourism, which is one of the pillars of the country's economy.[24]

The effect on skiing and sports-related tourism sounds like the impact on Colorado and the Aspen, Colorado, article that I referenced. Coming down the mountain from Coma Pedrosa's summit I had plenty of time to think. I thought about my lifelong love for the mountains, the crisp mountain air, pure water in its creeks and rivers, and its deep snow.

This love for mountains is a core value of mine, and it is a family value. Wallace Stegner, American novelist and environmentalist—who was often called "The Dean of Western Writers"—used this expression: "Retreat upward!" To me, that means: "Go to the mountains!"

David McKenna, two-time college president and seminary president—at age 28 the nation's youngest college president—taught us this truth about retreats that I quote quite often:

> *You go to the beach and the ocean to get your life in rhythm with the tides and*
>
> *You go to the mountains to get a better, longer view and vision for your life.*

Returning from the summit, I saw Scott waiting at the TH. Lou Ann and Savanah were close by having pizza for lunch at Surf Arinsal, a local restaurant serving Argentinean cuisine. My brother Rick had called from Arkansas while I was on the mountain and relayed the news that my dad had died peacefully in his sleep late Friday night. That was our Saturday morning, and I had felt Dad's presence with me several times on the mountain. *I was not alone.*

Dad had celebrated his 96th birthday just eight days earlier and privately we had talked about the mountain. He told me how much he loved me and my family and the great relationships and great adventures we had. Dad showed us how to live and he showed us how to die. *This summit is for you Dad!*

That evening while the girls visited and Scott watched a soccer championship game between *Real Madrid and Atletico de*

Madrid, I planned Dad's memorial service knowing his favorite Bible verses and Christian hymns. I knew we would sing *How Great Thou Art* which includes the memorable stanza "When I look down from lofty mountain grandeur and hear the brook and feel the gentle breeze."

I also knew that I would speak on the scriptures that say, "I have fought the good fight, I have finished the race, I have kept the faith." *Dad was part of the Greatest Generation born between 1901 and 1924. They don't make them like that anymore!*

On Sunday morning we began our return journey to Barcelona. We stopped along the mountain road at Organya, walked through a shop or two, and bought an espresso. The views were inspiring. As we travelled, we could see "Saw Mountain" in the distance and the memories of Santa Maria de Montserrat, and its peaceful cloister comforted my soul. Back in Barcelona with Scott and Savanah's host family, Mateo-Roca, with whom they were living, we were served paella with fresh fish they caught at the coast! And, of course, a glass of muscatel for dessert!

Well, so much for clear skies! It was raining hard and heavily in the Barcelona area on Monday and Albert (the dad) drove us to the airport around noon for our three-hour flight to Tenerife in the Canary Islands. Lou Ann was excited about an island adventure with warm beaches, and I had another climb in mind. We walked around Santa Cruz (a very industrial city) and had a wonderful dinner at El Lateral 27. Lou Ann is not a fish eater and so she ordered chicken while I had the excellent cod fish.

On Tuesday after many attempts to secure a permit to climb Mt. Teide and overnight at Altavista hut, the concierge named Salvador encouraged us to rent a car and drive to the Parque Nacional for the cable car to the top. We did that and hiked some trails, but without a permit, you cannot hike up to the volcano's crater. That was a little disappointing but there were many people at this obvious tourist attraction. Mt. Teide is a World Heritage Site.

We returned to the city by a coastal route of La Orotava and Puerta de Cruz. This was a lush area with lots of beautiful flowers and thick vegetation. We spent six days on the Canary

Coma Pedrosa, Andorra, 2014

Islands—walking and sightseeing, enjoying a spa day, swimming in the pool, shopping, and dining out. We found a few favorite restaurants—Café Atlantico was one—but generally we followed our meal practice of buying cheese, bread, and olives at a market and eating on the run or in our hotel room.

I didn't do a good job planning this extra excursion, so we didn't find the quiet and isolated scenic beaches that Lou Ann was hoping for on the Canary Islands. *Sometimes I am just too focused on climbing mountains.* Our route home took us back to Barcelona for one night with the family there and then our direct flight next day to Atlanta. This was a great ten-day trip, and I recorded another country's highpoint!

Later in 2014 my brother Rick and I would climb Wheeler Peak in New Mexico on August 25th to scatter Dad's ashes from the summit in his memory and consistent with his wishes. Dad shaped us and led us by example. He was a practical *and* a visionary guy. This quote from Jean Grou describes some of his influence:

> To direct a soul is to lead it in the ways of God; it is to teach the soul to listen for the Divine inspiration and to respond.

Dad was always planning another travel adventure. Since mom and dad were not rich, in our formative years the trips were usually camping in the mountains or exploring the pueblos and art galleries of northern New Mexico. As we grew into young adulthood, the travel expanded to Philmont Scout Ranch in New Mexico and the National Scout Jamboree in Colorado Springs, Colorado.

Summer vacations took us to the New York World's Fair, Niagara Falls, Oak Ridge, Tennessee, in the Smoky Mountains where mom dad met and married, the Grand Tetons in Wyoming, Yellowstone, the 1984 Summer Olympics in Los Angeles, the Grand Canyon, and many more. *It is a legacy we continue.*

By early September 2014, I was back in Colorado to do some solo climbing in the Collegiate Peaks area—Mt. Yale, Mt. Princeton, and Mt. Harvard. This was my journal entry of September

5th: "My good friend Steve is in the UK for what he calls 'study and sabbath.'

I am traveling to Colorado for my annual 14ers trip. I call it 'mountain research' and I find restoration in the mountains and wild places. My field research focuses on wilderness, the human need for it, and the Divine design in nature." It is research inspired by Dad.

Wilderness is explained as an uninhabited, uncultivated, and inhospitable region, a wild place. Edward Abbey called wilderness "a necessity of the human spirit." In his book Last Child in the Woods, author Richard Louv writes, *through nature, the human species is introduced to transcendence, in the sense that there is something more going on than the individual. Most people are either awakened to or strengthened in their spiritual journey by experiences in the natural world.*

8

Chugach Peaks, Alaska, 2015

Coastal Erosion and Pipeline Construction

Adventure is worthwhile in itself.
—Amelia Earhart

After returning to Colorado in September 2014 alone to climb Mt. Harvard from the North Cottonwood Trailhead, and in late September 2015 alone to climb Mt. Columbia and Mt. Antero in the Sawatch Range, I began to pursue more solo climbs. I soloed Mt. Shasta in July 2009 after a week-long retreat in Boulder Creek, CA. That was an awesome experience!

Read the Mount Shasta chapter in my book *Retreat Upward: A Mountain Pathway for the Soul* (Wifp and Stock, 2024). Here are three *observations* that I recorded on Shasta:

1. I climb to know myself, to explore my limits, to exercise good judgement, and to restore my soul.
2. Climbing solo is fulfilling but a climb is usually better shared.

3. Several times during the Shasta ascent on summit day, I had to stir up the desire and strength to go on.

On September 30, 2015, Lou Ann and I boarded a flight to Alaska where she would be attending an International Economic Development Conference in Anchorage, and I would do some climbing. That travel day was a long one from Little Rock to Chicago to Seattle to Anchorage. We had great flight connections and no flight delays. The return flight would be direct from Anchorage to Chicago overnight. Traveling with a purpose is awesome! I've hiked and climbed on all seven Continents, and I'm always amazed at how quickly and efficently you can be somewhere else in the world.

Lou Ann and I planned two days in the Anchorage area for sightseeing and exploring before the IEDC-Alaska train excursion to Seward. We went to find breakfast downtown on the first day and discovered the Snow City Café that boasted a great breakfast (President Obama had eaten there recently.) I had salmon cakes with eggs and potatoes and Lou Ann had a breakfast burrito with sausage, though a bit pricey.

Despite the overcast skies and drizzle, we walked all the downtown streets, shopped for a few souvenirs, and decided to find a grocery market since prices seemed high in local restaurants. The City Market Grocery was a long walk out of the Central Business District at Avenue I and 13th Street, but it was a quality market where we could buy a few snacks and grocery items. That evening we did return downtown for dinner at the Glacier Brewhouse Restaurant. I remember having delicious fish and chips while Lou Ann chose roasted chicken!

On our second day the rain continued in a steady drizzle that reminded us of Seattle and we kept walking and exploring. We toured the Anchorage Museum with great historical exhibits on native people's groups and in the afternoon found a large shopping mall. I tried some salmon quesadilla and polenta there.

We did a lot of walking ignoring the rain, drizzle, and dreary weather. Friday evening the city was promoting "First Fridays" when retailers stayed open late and most served finger foods and

Chugach Peaks, Alaska, 2015

snacks. We found the Aurora Art studio and lingered for a while, viewing some unique collections of Native and Alaskan art.

Saturday was going to be a very long day taking the Alaska Railroad tour to Seward, the Kenai Fjord boat cruising the bay and inlets, and returning by train to Anchorage hopefully before midnight. I love train travel, and the Alaskan train was no exception. The views from the dome cars and open-air cars were exhilarating despite the overcast skies. Breakfast on the train was an exceptional culinary experience of reindeer sausage, eggs, potatoes, and coffee.

We saw very little wildlife from the train on its 4 to 5-hour journey but occasionally a single bear or moose would be spotted along a wide riverbed. There are just so many mountains, rivers, and lakes in this wide expanse of wilderness from the Cook Inlet up the Turnagain Arm to the coast at Resurrection Bay. The Bay received its name from Alexandr Baranov, who was forced to retreat into the bay during a bad storm in the Gulf of Alaska. When the storm settled it was Easter Sunday, so the bay and nearby Resurrection River were named in honor of it.

This paragraph from the Alaska Railroad brochure caught my eye:

> *At more than twice the size of Texas, the vast Alaskan wilderness has much to offer the adventurous outdoor traveler. Whether you choose to kayak the Kenai River or hike Denali National Park, natural wonders are abundant. But with its proximity to the fast-thawing Arctic, Alaska is already experiencing major changes in the form of coastal erosion, sea ice retreat, and permafrost melt. The state's many ice caps are receding at extraordinary rates, triggering landslides so intense they have registered on the Richter scale. Another devastating effect of higher temperatures is wildfires, which have destroyed more of Alaska's forest in the past decade than any previous. The number of wildfires is expected to double by 2050.*[25]

Alaska's vastness is obvious but one must look much closer and clearer to *observe* the damage of climate change. *My eyes were straining to see what I could see.*

Moving Mountains

We had chicken Caesar wraps and hot drinks for lunch as we boarded the Kenai Fjords Tour ship. This was a premium six-hour cruise on which we saw humpback whales, sea lions, sea otters, many species of birds including puffins, Dall porpoises, orcas, and even one black bear on the shoreline.

The tidewater glaciers and sheer number of massive glaciers meeting the sea was almost more than you could take in. The Aialik Glacier was one mile wide and 400 feet tall! One of the largest glaciers up on the Harding Ice Field was 5000 feet tall. We slowed and watched for an hour or more the glaciers calving huge blocks of ice into the sea.

> A glacier is a large mass of snow and ice that has accumulated over many years and is present year-round. A glacier flows naturally like a river, only much more slowly. At higher elevations, glaciers accumulate snow, which eventually becomes compressed into ice. At lower elevations, the "river" of ice naturally loses mass because of melting and ice breaking off and floating away (iceberg calving) if the glacier ends in a lake or the ocean. When melting and calving are exactly balanced by new snow accumulation, a glacier is in equilibrium and its mass will neither increase nor decrease.
>
> Alaska has hundreds of named glaciers at high elevations and at low elevations. On average, glaciers worldwide have been losing mass since at least the 1970s, which in turn has contributed to observed changes in sea level. Alaskan glaciers are no exception. A longer measurement record from a smaller number of glaciers suggests that they have been shrinking since the 1940s. The rate at which glaciers are losing mass appears to have accelerated over roughly the last decade.[26]

Leaving the boat, we did some walking from the Marina, around the town of Seward, and back to the train station. We had a great dinner of roast beef, potatoes, and green beans in the train's dining car on our return trip. I slept some on the train after dark (8:00 pm) and we arrived back in Anchorage around 11 pm. This

Chugach Peaks, Alaska, 2015

was an expansive learning experience as we filled the day with Alaskan adventures!

We slept a little later the next morning. Lou Ann's conference was scheduled to begin after lunch, and I was looking for a way to get up into the mountains. For breakfast, we split the excellent eggs benedict at Sack's. We did more walking and found the Convention Center so Lou Ann could pick up her conference materials. We sat down and reviewed her sessions and thought through our options. We returned to the Glacier Brewhouse for a great lunch at this quality restaurant.

I rented a car and drove Lou Ann by the Convention Center for the Plenary Session and then drove up to the Chugach Mountains north of Anchorage. The Chugach Mountain range of southern Alaska is about 250 miles long and 60 miles wide, extending from the Knik and Turnagain Arms of the Cook Inlet on the west to Bering Glacier and Tana Glacier on the east. Its position along the Gulf of Alaska ensures more snowfall in the Chugach than anywhere else in the world, an annual average of over 800 inches.

I had chosen Flattop Mountain (3510 feet) for its proximity to Anchorage for my afternoon adventure. I parked at the Glen Arms TH to begin my hike and climb on the Flattop Mountain Trail. The trail was first slush and then snow-covered on most of the approach. The distance was shown as three miles roundtrip with a gain of 1250 feet.

The fresh snow was 10–12 inches deep on the higher slopes and the crux was a steep trough with about 18–20 inches of new snow. I had failed to bring my crampons or Yaktrax, so I climbed carefully using my hands and kicked steps for solid toeholds. Soon I was alone on the summit of Flattop! I didn't see anyone going up or coming down.

A cloud cap had formed and covered the summit, so I only stayed on the top for maybe ten minutes. The descent required me to back off the top and down climb the trough slowly. It was a great snow climb. I ascended in 75 minutes and descended in 70 minutes between 2:50 and 5:20 pm. I drove back to the hotel, met Lou Ann in the room, and we had cheese, bread, and fruit for our evening meal. We decided to stay in our room and rest up for tomorrow's adventure. Lou Ann has a light schedule, and we may take a drive somewhere while I have the rental car.

We woke up to dense fog outside which was supposed to burn off by late morning. We talked about driving up to Talkeetna but cancelled that plan. Instead, Lou Ann would find some sessions that sounded interesting at this ultimate stage in her career, and I would climb some more mountains in the Chugach. I ate a big breakfast at the hotel and then drove Lou Ann to the Convention Center about 10:30 am. I went to the Anchorage REI store and bought some Yaktrax ("snow chains") for my Asolo boots and then headed to the mountains.

I parked at the Prospect Heights TH about 11:30 am. Signs warned of bears on the trail but all I saw was a moose cow and her calf. I took the Wolverine Bowl Trail around 11:45 am to hike up to Wolverine Peak (4492 feet) and summited at 2:30 pm. I hiked over to another peak called Near Point (3500 feet) and then returned to the parking lot by 5:15 pm covering ten miles total.

I drove back to the hotel by 6:00 pm and cleaned up for a nice dinner with my wife. We watched a beautiful sunset overlooking Cook Inlet at Simon and Schubert's restaurant. What a spectacular view and sensational food! I had a seafood salad with salmon, shrimp prawns, crab meat, artichoke, asparagus, tomatoes, and deviled eggs. Lou Ann had a chicken spinach salad. *What a great way to close out a great day!*

The next day we got an earlier start as I had decided to drive north to Talkeetna for a long look at Denali. There was dense fog again, but the traffic moved steadily all the way to Wasilla. I knew some people there but thought I might stop on the way back. I made the drive to Talkeetna in two hours, arriving there just after 10:00 am. The first thing I noticed was the airplane hangars and flight services I had always heard about—K2 Aviation, Talkeetna Air Taxi, Fly Denali, Sheldon Air Service!

I walked around the small town (population of about 900), took some photos, and stared at Denali in the distance about 75 miles away. It still looks big! It was a bluebird day in early October, and it was a great view of the highest mountain in North America! I walked about two miles in Talkeetna, shopped for a souvenir or two, and bought some coffee at the Conscious Coffee house. I still have a desire to climb Denali even with all the snow and wind!

> Denali (also known as Mt. McKinley, its former official name) is the highest mountain peak in North America with a summit elevation of 20,320 feet. Denali is the third most prominent and third most isolated peak on Earth, after Mt. Everest in Nepal and Aconcagua in Argentina. It is one of the coldest mountains in the world with temperatures during the spring and summer climbing season of as low as -40 degrees F and with winds of 80 to 100 mph lasting several days. Technically, the climb is not so difficult, but low temperatures and awesome winds conspire to make this mountain environment one of the harshest on earth.
>
> Although several thousand climbers have reached the summit, many have been killed and many have required rescue. Avalanches, crevasses, and altitude sickness can all prove fatal. Frostbite is the most common malady experienced. The usual climbing season for Denali is May through mid-July to avoid the cold storms and high winds of winter and the increased crevasse danger of late summer. The first ascent (FA) of Denali occurred on June 7, 1913, by Hudson Stuck, Harry Karstens, Walter Harper, and Robert Tatum.[27]

I left Talkeetna around noon and drove back to Wasilla to look up a friend that I knew. I found some employees in his office, but he was travelling. I got back to the hotel in Anchorage about 2:30 pm. This was a very nice drive in Alaska, and a single-day trip with mostly sunshine north of Wasilla. We attended an IEDC reception at the Anchorage Museum in the evening with more Alaskan cuisine—reindeer stew, shrimp pasta, salmon sushi, and specialty cheeses. I enjoyed meeting some of my wife's colleagues in economic development and making new friends from many places.

Our last day in Anchorage was going to be a long one since our "red eye" flight for Chicago did not leave until 11:30 pm. The weather forecast called for "rainy and cooler temps." Since we had the rental car, we filled our day with more sightseeing in and around Anchorage, caught a movie called "The Intern" in the afternoon, and found an Olive Garden restaurant for a nostalgic dinner. When the kids were growing up, we dined at an Olive Garden restaurant often.

Despite the rainy and gloomy weather that we had experienced, we had a positive impression of Anchorage. Here is a little more history:

> *The Anchorage city limits span 1961 square miles encompassing the urban core, a joint military base, several outlying communities and almost all of Chugach State Park. The Ted Stevens Anchorage International Airport is a common refueling stop for international cargo flights and a home to a major FedEx hub, which the company calls a "critical part" of its global network of services. Alaska became an organized incorporated U.S. territory in 1912. Anchorage, unlike every other large town in Alaska south of the Brooks Range, was neither a fishing nor a mining camp. Anchorage was incorporated in November 1920.*
>
> *Construction of the Alaska Railroad became the main economic driver and began to draw more transportation services with air transportation and the military. Anchorage International Airport was opened in 1951. Elmendorf AFB and the U.S. Army's Fort Richardson were constructed in the 1940s and served as the city's primary economic engine until the 1968 Prudhoe Bay discovery on the Alaska North Slope shifted the thrust of the economy toward the oil industry. The resulting boom in 1975 spurred further growth in Anchorage. From the Trans-Alaska Pipeline to drilling in the Alaska National Wildlife Refuge to the Exxon Valdez oil spill in Prince William Sound in 1989, the state has not been exempted from environmental issues.*[28]

The Biden Administration announced plans in January 2021 for a temporary moratorium on oil and gas leasing in Alaska's

Chugach Peaks, Alaska, 2015

Arctic National Wildlife Refuge (ANWR) after President Trump issued leases in a part of the refuge considered sacred by the Gwich'in people. ANWR comprises 19 million acres of the north Alaskan coast between the Beaufort Sea to the north, Brooks Range to the south, and Prudhoe Bay to the west. Much of the debate over whether to drill in the 1002 area of ANWR rests on the amount of economically recoverable oil, as it relates to world oil markets, weighed against the potential harm oil exploration might have upon the natural wildlife, particularly the calving ground of the Porcupine Caribou.

Both of Alaska's U.S. Senators, Republicans Lisa Murkowski and Dan Sullivan, have indicated they support ANWR drilling, as does President-elect Trump. The Alaskan Inter-Tribal Council, which represents 229 Native Alaskan tribes officially opposes any development in ANWR. The Gwich'in tribe adamantly believes that drilling in ANWR would have serious negative effects on the caribou herd that they partially rely on for food.

The primary effects of climate change focus on impacts on wildlife and the natural resources like water and food. The climate change issue with respect to ANWR drilling for oil and gas revolves around further dependence on fossil fuels for energy. As has been referenced several times, the urgent need to reduce greenhouse gases and carbon emissions requires a shift to renewable and more sustainable energy sources.

These are harsh lessons on the environment and the economy. I want to learn all I can. One of my self-published collections of poems is called Mountains, Hills, and Prairies. One of those poems that I wrote in the 1970s is simply called "Alaska."

Alaska

Rich men take the land, crush what stands in their way,
The beauty may forever die, one minute it's here and then it's gone,
Their promises all turn to lies.
Well, these are the mountains,
And they are my home,
They shelter the elk and the deer,
With dollars in your eyes,

Moving Mountains

Some developer's disguise,
Your money is not welcome here.
Now listen you miners and you oil men too
Listen you men that build roads,
My grandfather spoke of a time before fences, of endless forests and trees,
We've got to have wilderness,
If we want to be free.
Be careful Alaska, what they did down in the states surely they will do here.

9

Snowdon, Scafell Pike, Ben Nevis, 2017

Carbon Emissions

Life was meant for good friends and great adventures.

In 2016 I hiked seventeen smaller summits and Ozark Highlands highpoints, but the only challenging mountains were a snow climb of Wheeler Peak, NM, (13,161 ft.) in May and a climb of Culebra Peak, CO, (14,047 ft.) in June. Get outside!

Lou Ann and I had a wonderful trip to New Zealand (probably our favorite travel spot ever) in late October and early November 2016. My brother Rick called me as we were driving across New Zealand's South Island and told me that our mom had died on the evening of November 6th at the nursing home in Arkansas.

Mom was ninety-eight and half years old having been born under a tree east of the Sandia Mountain Range of New Mexico in May 1918. She was the second child of ten children born to her parents, Rella Cleveland and Maude Valentine. She became a registered nurse and took her first nursing job at the base hospital in Oak Ridge, Tennessee, where she met a Texas man named Joe Nisbett. Rick and I decided that we would hold a memorial service with family at Christmas.

Early in 2017, my wife and I received an invitation to an Adventure with Friends to spend a week at Oxford, England, studying the life and works of C. S. Lewis and other Inklings authors. *You don't find adventure, adventure finds you.* Said another way in one of the lessons from the Hobbit stories: *"When adventure comes knocking, let it in."* My Jesuit friend Rick Ganz says, *"Faith is a highly developed capacity for adventure."*

We accepted the invitation to spend a week at Oxford in late August 2017. We would journey with a few close friends and twenty or so new friends eager to learn. We heard knowledgeable speakers discuss C. S. Lewis and J. R. R. Tolkien and show us the hearts of these great men who had faith. Oxford truly is the preeminent storehouse of knowledge and the seat of intellectual ideas and thought.

One of the Oxford apologetics speakers used this expression, "Helping thinkers believe and believers think!" It would probably bore the reader for me to recount all the lectures we heard but I do describe some of the trip highlights below.

After my son-in-law Scott and I climbed Carrauntoohil in Ireland, we began to talk about hiking to the highpoints of England, Scotland, and Wales. Since Lou Ann would fly home and return to work after the week at Oxford, it just seemed logical for Scott to fly to London on that day and meet me in Oxford.

Scott is a brilliant guy with a master's degree in Exercise Physiology and we enjoy our adventures together! I will describe our UK adventure and identify climate change issues as we go along. First, I promised some Oxford highlights.

Lou Ann and I have reflected so many times over the past three years on the Adventure with Friends with the purpose of personal growth and faith-building. The trip easily ranks as one of our most meaningful travel experiences and we have chosen to travel to many places. *We are ambassadors; God is making his appeal through us. We speak for Christ when we plead, "Come back to God."*

Lou Ann was a member of the Texas Girls' Choir; a touring choir founded by her Aunt Shirley Carter and sang with them for ten years. The TGC were ambassadors through music and traveled domestically and internationally every year. Lou Ann has been to some countries I still haven't visited—Israel, Egypt, Italy, Germany among others.

My priorities of climbing mountains and public speaking have taken me on many travel adventures. I have been to some countries that my wife has yet to visit, India, Kenya, Tanzania, Liberia among others. It may sound like we are rich and famous, but we are not. Both sets of our parents thought that travel was important and took us along. *Travel is a family priority.*

We did the same with our children and they are doing the same with our grandchildren in that tradition. All people spend their money in certain ways and live out their own priorities. *"All we must decide is what to do with the time that is given us."*

Our Oxford Highlights:

1. It was astonishing to lodge in the Malmaison Oxford Hotel practically on the campus and itself a former prison. "Malmaison" is French for "bad house." Oh, if walls could talk!

2. It was a unique opportunity to hear speakers on topics ranging from the Inklings, including Lewis and Tolkien, to Falconry. We heard Oxford dons and graduate students pontificate on literature, philosophy, religion, history, and apologetics. I wrote down so many quotes!

3. We had the rare privilege of interviewing and hearing the remarks of Walter Hooper, the personal secretary of C. S. Lewis, for many years. Lewis died the same day as John F. Kennedy on November 23, 1963, and yet we could hear Walter at age 86 recalling so many details and personal insights about Lewis. "Walter Hooper was himself a very bright man, with a gift for writing, and a set of clear, strongly felt principles." Walter Hooper died on December 7, 2020.

4. It was a highlight to walk the streets of Oxford where so many remarkable people and world leaders have walked since 1096. Oxford is the oldest university in the English-speaking world and developed rapidly after 1167 A.D. The University is a collection of individual colleges and a walking tour of them was an education as well. We had our Adventure Celebration Banquet in the Hall upstairs about Vault and Gardens at St. Mary's College.

5. It was an amazing invitation that my wife and I received one Monday morning to sit in on the daily world briefing at Oxford Analytica. We were not allowed to speak nor ask questions as regional, and country experts updated the team on geopolitical and economic events throughout the world. Thanks Maegan Moore!

6. Heading in a small boat to the Eastern Shore after the death of Boromir, Frodo tells Sam (his companion and best friend): "I am going on alone." To which Sam replies, "Sure you are! And I am going with you!" The friends we dined with, laughed with, raised a glass to at The Eagle and Child pub on St. Giles' Street, and grew to know better was an Adventure with Friends highlight for sure.

7. My last journal entry before leaving Oxford says, "To reach the City of God we must leave the City of Man." That may refer to our physical death, but I think it applies to walking and talking with our friends like the Inklings did along the dirt paths and into the meadows beyond the busyness of the city of Oxford. I know God must visit the cities of mankind but I'm not sure he lives there. I find God in the mountains and the wild places where I hike and climb . . .

On September 2, 2017, Scott and I rented a car and drove five hours from Oxford to Rhyd Ddu, Wales, where we had reserved a bunkhouse near Cwellyn Arms pub. Rhyd-Ddu is a small village in Snowdonia, North Wales, which is a starting point for hikes up Snowdon via Moel Hebog, Yr Aran, and Nantile Ridge. I have some Welsh heritage through my Grandfather Valentine and

Snowdon, Scafell Pike, Ben Nevis, 2017

have long wanted to travel Wales. Visiting with some local Welsh folks, they recommended that we do not even try to pronounce the Welsh words!

We ate breakfast at Cwellyn Arms at 8:00 am and waited in our converted barn bunkhouse for the rain to let up. At about 10:30 am we drove to the Car Park near the Snowdon Ranger Path TH. We left the TH at 10:45 am on what the guidebook said was an eight mile and six-hour roundtrip hike from 490 feet to 3560 feet at the summit, a gain of 3070 feet. Despite the lower elevations on these mountains, the vertical gain of 3000 or more feet is like our Colorado climbs though their higher altitudes make a big difference.

We had rain, blowing rain, and high winds of 40–50 mph under foggy skies for all sections of this hike. We had researched and anticipated the UK weather at this time of year and wore our best mountain raingear.

Quoting a famous UK guy, Bear Grylls: "If it ain't raining, it ain't training."

We ascended to the top of Snowdon in 2 hours, 5 minutes, and only spent 10 minutes on top. With no summit view to speak of, we left the top at 1:00 pm and descended to the Car Park lot by 2:45 pm. That was a four-hour round trip!

We returned to the bunkhouse, built a fire, and hung our wet clothes on some racks for drying. Many of the trail lodging places have dry rooms for wet clothes but our bunkhouse fireplace did the trick. We headed to Cwellyn Arms for homemade soup and ale. These UK pubs, short for "public house," and Country Inns are warm and rustic gathering places where you can eat Welsh lamb, trout, burgers, or fish pie. We stayed in the bunkhouse for a second night.

With all the rain, which is normal here, I wanted to understand climate change issues in the U.K. in general so here is an overview. First, I discovered a story from Wales in particular.

> *Aberystwyth is a seaside town which has already had experiences of extreme weather. Five years ago, properties were evacuated as waves crashed onto the promenade. It seems*

> as good a place as any to take a look at what impact climate change may be having on communities in Wales. Its university is also the base for scientists who are researching the wider climate issues—and how we might adapt.
>
> Alun Williams, cabinet member for carbon management in Ceredigion, said rising sea levels were threatening its flood defences. It is an example of how global warming is already affecting Wales—the country's 1680-mile-long coastline is on the frontline! "We need to manage this on a Wales-wide level—you can't put all the cost on seaside towns and county councils—we need a wider strategy." Williams said.
>
> The changing climate is also showing signs of hitting Wales' wildlife and the habitats they rely on. "We're losing a lot of the species which have been here for millennia—egrets, black grouse, golden plover—and that's really sad as many of them are globally quite restricted." (Williams) Farmers are facing challenges too, with predictions Wales will face more rain, warmer temperatures, and fiercer storms in the future. "Tackling climate change and species extinction are not issues which can be left to individuals or to the free market. They require collective action, and the government has a central role."[29]

Scott and I drove 5 or 6 hours today making stops in Caerfarnon for coffee and pastries at a Costa café; at Rhyl to walk alone the coast of the Irish Sea; and at Lancaster for fuel and lunch. We arrived at the Wasdale area on Wast Lake about 4:00 pm. We had the pea and ham soup and ale at Ritson's Bar at the Wasdale Head Inn. I love this Inn and Pub which sit at the end of the road approaching England's highest mountain, Scafell Pike (3209 feet).

We had a reservation at the Murt Barn loft which was about four miles back to the south on the road we came in on. The Barn Loft had eight mattresses on the hard wooden floor where we put our sleeping bags. We turned out to be the only people there that night. We had lots of rain today across central England, but we planned to climb Scafell Pike the next morning.

Snowdon, Scafell Pike, Ben Nevis, 2017

We had some coffee and snacks at the National Trust Car Park this morning and then left the Wasdale Head TH around 10:20 am with rainy and foggy conditions. The air temperature was very warm, and the air was humid. There were many individuals and groups "tramping" (trekking) up the Scafell Pike trail today!

Scott and I moved fast, and we passed most of the hikers who had left the TH ahead of us. We reached the summit at 12:15 pm, just under two hours in weather with low visibility, windy, and rainy. No reason to stay on top! We descended between 12:20 pm and 2:15 pm on wetter, slippery trails.

We ate an early and large meal at the Wasdale Inn and then headed for Scotland, arriving in the town of Gretna by 6:00 pm. This was the only place we had not reserved a room, but we found one at the Gretna Hall Hotel in Gretna Green, Scotland. We learned later that Gretna Hall is a historic marriage house. "Graitney Hall" as it has been known in the past has a rich history of ownership.

It was built in 1710 as the family mansion and the estate belonged to the Johnstone family. The family coat of arms is engraved in stone displayed still above the entrance. Scott and I had a good laugh about staying together in a marriage house! We did have a good night's rest and a great breakfast there.

We drove the two or so hours bypassing Glasgow, north to Stirling. We spent several hours walking the town and visiting Stirling Castle in this historic city of Robert the Bruce and William Wallace—scenes right out of Braveheart. Well, you can imagine it! We drove several more hours past Glen Coe and on up to Fort William. We stopped there to walk some more and to buy groceries. Our hostel was just above Fort William at the Ben Nevis Inn on the Achintee Farm near the Ben Nevis Trailhead.

After unloading our gear and getting settled in this modern lodge, we hiked up the trail for about a mile and back down in this spectacular setting at the base of the mountain. The weather forecast for the next day was stormy and rainy, so we decided to explore the area and hope for better climbing weather in two days.

A low-pressure system established itself in the area on Thursday as forecasted and rains fell, and the wind blew. Highland weather! We made a breakfast of baguette French toast and coffee this morning. We walked down into town to the Ben Nevis Distillery for a tour and a tasting of their scotch whiskey made there since 1820. I am not a whiskey drinker, but the distilling process was interesting, and the tour was worth the price.

After the whiskey tour, we walked around the shops and met hikers who told us all about the extensive trail system in the UK. One woman from St. Louis had hiked the 100-mile Highlands Trail up from Glasgow in about eight days. She told us about the West Highlands Trail and an East Highlands Trail among many others in Scotland. This woman had hiked the 550-mile Camino de Santiago trail in Spain and a 400-mile trail from northern Italy to Rome. That made us hungry, so we made some sandwiches for lunch.

In the afternoon we drove to the end of the Glen Nevis Road to park and hike 4+ miles from the lower falls and up to the alpine meadow at the base of Steall Falls. We had to give the three-cable bridge across the river a try. That was fun! This is a longer approach and route on Ben Nevis. We couldn't see Ben Nevis today due to the fog and rain-heavy clouds. This trail continues to the Speann Bridge (19 miles), the Ring of Steall to Ben Nevis, and other Munros—peaks 3000 feet and taller.

Back at the Ben Nevis Inn and Hostel we made pasta with chicken and pesto plus garlic bread sticks for supper. This was a good day despite the ever-present rain and low clouds! Here is the UK climate change general information I mentioned:

> *Climate change in the United Kingdom (UK) has been a subject of protest and controversies and various policies have been developed to mitigate its effects. One goal is to decrease greenhouse gas emissions in the UK by 50% on 1990 levels by 2025 and by 100% on 1990 levels by 2050. In December 2020, Boris Johnson declared that the UK will set a target of 68% reduction in GHG emissions by the year 2030 and include this target in its commitments in the Paris Climate Agreement.*

Snowdon, Scafell Pike, Ben Nevis, 2017

> *The British government and the economist Nicholas Stern published the Stern Review on the "Economics of Climate Change" in 2006. The report states that climate change is the greatest and widest-ranging market failure ever seen, presenting a unique challenge for economics. The Stern Review provides prescriptions including environmental or carbon taxes to minimize economic and social disruptions.*
>
> *William Nordhaus, an economist from Yale, said that the Stern Review should be read primarily as a document that is political in nature and has advocacy as its purpose. This assessment seems to be justified by this statement within the Review itself: 'Much of public policy is actually about changing attitudes.'[30]*

Well, it has been a wet week in the British Isles for us!

> *By 2014 the UK's seven warmest years on record and four out its five wettest years had occurred between the years of 2000 to 2014. Higher temperatures increase evaporation and consequentially rainfall. In 2014 England recorded its wettest winter in over 250 years with widespread flooding.[31]*

We had more rain today but resolved to climb. We made eggs and toast for breakfast, dressed in layers for the climb, and hit the trail at the Achintee TH at 8:15 am. We easily summited Ben Nevis— the highest mountain in the UK—at 11:00 am. We spent about 15 minutes on top, took pictures, and shared a sip of scotch whiskey offered by a Scottish climber and friends.

We told them we came from Texas and the climber laughed as he said, "Soooo, shall we start the summit celebration by sacrificing the Texans now?" Not sure if they were serious or if that was the price of their precious scotch whiskey, we turned to leave and descended faster than we planned! We did stop and talk to people on the trail as many groups were tramping the trail today. Ben Nevis was climbed and recorded as a first ascent (FA) by James Robertson on August 19, 1771.

Now, in full disclosure, I have Scottish blood and our Nis-bett name can mean either "nose-biter" or "none-better" in the Gaelic language. On my first trip to Scotland from France and England in 1966, I found the Scottish people we met in Edinburgh and Glasgow to be the friendliest people I had ever met. This trip was no different and we had rewarding talks with perfect strangers on the trails and in the pubs.

It had been foggy with rain and drizzle falling all day. We got back to the TH at 2:00 pm. As we compared our rainy hikes on the three UK mountains, Ben Nevis sits at a higher latitude and rises to a higher altitude and therefore was a little colder on top. Our view from Ben Nevis was also obscured by clouds but descending just a little lower we enjoyed spectacular views of the Fort William area, its lochs, and its valleys. Another Hobbit quote seems to fit: *Find the enchantment all around you!*

Scott and I finished the Three Peaks Challenge today after summiting Snowdon, Scafell Pike, and Ben Nevis (4406 feet). We learned that the Challenge is a real "thing" in the UK. It took us a week but get this:

The National Three Peaks Challenge involves climbing the three highest peaks of Scotland, England, and Wales within 24 hours. The total walking distance is 23 miles, and the total ascent is 10,052 feet.

Well, that is a lot of driving as well and we will just have to return! From the Snowdonia Range in Wales to the Lake District of England to the Scottish Highlands and Lochaber, we have had a wonderful trip in the British Isles. It has been rainy and wet as well as humid and windy while hiking this week with that cooler temp on Ben Nevis. What a great adventure! *When an adventure comes knocking, let it in!*

We left Fort William reluctantly the next morning at about 8:30 to drive to Glen Coe. At 9:15 am we just had to stop. There is so much in Scotland to take in. If I could just see the Isle of Skye, the Outer Hebrides, Iona, Inverness, Loch Lomond and Loch Ness, and The Borders where the Nisbetts came from, I would be happy. We took a hike up into the hills on the Three Sisters trail for several miles to get one more panoramic view. Groups of rock climbers with ropes and harnesses were going higher to climb the rock formations there.

Snowdon, Scafell Pike, Ben Nevis, 2017

We drove an alternative route to have a roadside stop in Corrander for lunch. Scott and Savanah had stopped here with his parents in 2012 touring around Scotland. In the convenience store there I happened to see and buy a Romney's Chocolate Covered Kendal Mint Cake. The candy bar was made by George Romney Ltd, and some were carried to the top of Mt. Everest in May 1953 on the first successful climb to the summit. I didn't eat the chocolate bar until I returned home to the States, and it was delicious and nutritious!

Scott had reserved a hotel room at a special price near the Glasgow Airport, the Erskine Bridge Hotel which had the necessary swimming pool, hot tub, and sauna. We checked in there and then drove the twelve miles back into Glasgow to walk around the University of Glasgow campus. Lots of students and families were strolling through the parks and green spaces there.

I found the Adam Smith School of Business on the campus and Scott took my picture as I recalled lessons from "the Founder and Father of Economics." Smith once told a friend that *he made it a rule when in company never to talk of what he understood.*

We had dinner back at the Hotel, relaxed in the pool, and rested up for our return flight the next day. Scott and I were building a great friendship and so we talked a great deal about many topics, some we understood and others not so much. Friendship is based on intentional communication and intellectual hospitality. Our conversation always included listening and asking great questions.

It would be nice to get home after travelling for 18 days. I call our mountain home in Arkansas "Base Camp." *Be it ever so humble, there is no place like home!*

10

Aconcagua, Argentina, 2018

Temperatures and Glacial Melt

Every few hundred feet the world changes.
—*Roberto Bolano*

After returning from my British Isles trip, I called Shawn and asked him to consider a trip to Argentina to attempt Mt. Aconcagua, one of the seven summits. Unlike other trips where my business could share the expenses or sponsors supported the cost of the trip, Aconcagua would be a pricey out-of-our-pockets adventure. In addition to the cost, we would be away from our jobs for 23–24 days.

The appeal of the climb was obvious to us but probably not so much to our wives. The word "expedition" means "to leave on foot" and this would be a three-week expeditionary climb for us, moving camps higher on the mountain until from high camp we would attempt to reach the summit at 22,800 feet. The trip is round. Watch for our return!

Shawn and I met in the Houston International Airport on January 1st to catch our overnight flight to Santiago, Chile. We then had a late morning flight on January 2nd over the Andes Mountain range into Mendoza, Argentina, arriving about 12:30 pm. We took a taxi from the Mendoza airport to the NH Hotel in the city center by 2:00 pm. We rested and unpacked our bags of gear. There was a lot of sorting gear to do before an inspection by our Acomara guides.

The NH Hotel -Mendoza Cordillera is a short walk from the city's main square, the best shopping streets, and sidewalk seating for dining out. Mendoza felt like a European city and was full of gourmet restaurants and sprawling vineyards famous for the Malbec grape. We found Giovanni's Italian café and enjoyed it so much that we had lunch and dinner there!

In the afternoon we met with Viviana with Acomara Guides— owned by Argentina Mountain Guides (AMG) and sold to Alpine Ascents International of Seattle later that year. We had to pay her our respective balances for the guided climb in cash and, with a currency exchange, those transactions and paperwork required a few hours. We arrived one day early for our climb which was beginning on Thursday, January 4th.

Miscommunications are common across culture and language barriers— but that's how we learn. On Wednesday, now a free day, we found one of the small sidewalk cafes that served breakfast and had coffee with a pastry and orange juice. We walked this interesting city of over one million people and hundreds of wineries. Two of the main industries of the Mendoza area are olive oil production and Argentine wine.

The region around Greater Mendoza is the largest wine-producing area in South America. Mendoza is one of the nine Great Wine Capitals of the world, and the city is an emerging wine tourism destination. So, in the afternoon, we took a five-hour complimentary wine tour of three wineries and an olive orchard with an olive factory and store. We grabbed a pizza from Zeus for dinner.

Thursday, January 4th, we did the same morning coffee and pastry routine because not many of the sidewalk cafes were open for breakfast. We walked more of the city and did some sightseeing and then had lunch at Il Panino. Mendoza is a great city!

Aconcagua, Argentina, 2018

In the afternoon we met our team of nine climbers and three guides. They gave us a general briefing and then came to each of our hotel rooms to carefully check all our gear and equipment. We had been provided a detailed gear list online, but each climber's gear was inspected as if our lives depended on it. Maybe they do!

We all walked together to a local Outdoor Outfitter to rent any items we didn't have. The only thing Shawn and I needed was a second Thermarest pad to provide more insulation from the cold ground. The rental prices seemed high, but you can't always bring everything on the plane and this rental lasted 21 days.

We would check out of the hotel early the next morning, leaving our personal travel bags in the hotel's secure room, and take a bus for a 3+hour drive on the Trans-Andean Highway to Los Penitentes. Before we left Mendoza, we stopped at a National Ranger Station to pick-up our wilderness permits and climbing documents.

In Los Penitentes we stayed overnight at the Ayelon Ski Lodge to acclimate in the foothills at 8000 feet elevation. We were served a huge dinner and then a big breakfast there on Saturday morning. We had a short drive to the park entrance but stopped on the way to unload all the heavy gear at Los Puquios depot which would be carried by mules to Base Camp at 14,435 feet.

At the Horcones TH, the team met with our head guide Pablo and his two assistant guides, Christian nicknamed "Soldier," and Javier, who Pablo sometimes called "Chan"—I assume after an actor and stuntman. "Soldier" had served in the military and liked discipline and motivating climbers which he would do on this climb.

In his physical and mental preparation for climbing Aconcagua, Shawn had hired a personal trainer to get him in top shape. From his days in the USNA and in submarine service, Shawn had always stayed in shape by running and lifting weights. We often climbed many mountains together and he was always fit and a strong climbing partner.

The personal trainer in Virginia put him through intense and repetitive workouts. I sent Shawn some of my exercise practices adopted from one of Ed Viesturs climbing books:

1. Repeats of stepping up and then down carrying dumbbells.
2. Balancing on one leg while squatting up and down.
3. Doing curls with weights.
4. Push-ups and Pull-ups.
5. Squats and Lunges.
6. Holding barbell weights by your fingertips.
7. Balancing on an exercise ball for core strength.

We entered the National Park, presented our $900 permits, and started up the Horcones Trail on the four-mile, three-hour trek to Confluencia Camp at about 10,000 feet. The hike was easy on a well-worn trail and carrying small personal backpacks. This was the only trail with a few trees, bushes, and green grass. We would spend two nights at Confluencia and do some acclimating hikes in the area.

We set up our North Face VE25 tents and worked as a team helping others with their tents. This was important for working as a team higher on the mountain. Long walks up and down and slowly acclimatizing to altitude is the best strategy for climbing high alpine peaks. AMG staff cooked hot meals at camp ideal for the stamina we would need with plenty of hot tea for hydrating well.

One of the staff cooked some scraps of meat one evening in the staff tent area for a very cautious but almost domesticated wild red fox. The man would place the meat on a stump and the fox would carefully approach the edge of camp, grab the meat with his teeth, and promptly run to bury it in the dirt some distance away.

We watched this game "play out" four or five times until the fox had buried each meat scrap in a different place. The human-animal connection looked like a nightly ritual or entertainment act. I left not knowing for sure. We didn't see much wildlife at all during the 21-day trip within the National Park.

On January 7th, we did a day-hike of about ten miles to Plaza Francia at about 12,500 feet and ate lunch beneath the spectacular South Face of Aconcagua. It was our first good view of the mountain. The first recorded ascent (FA) of Aconcagua was by Matthias

Aconcagua, Argentina, 2018

Zurbriggen in 1897, a Swiss mountaineer. He climbed throughout the Alps, the Andes, the Himalayas, and New Zealand.

The South Face is famous for its difficult ascents and some of the world's best climbers come there to test their skills on this 3000-foot perilous rock of hanging glaciers, snow chutes, and icy seracs. A French team first climbed the South Face in 1954 and other famous climbers including Reinhold Messner in 1974 have pioneered routes up the South Face. The French Team in 1954 descended the Normal Route on the mountain's north side after being caught in a raging storm and reaching the summit at 8:00 pm late evening. The six climbers returned with severe frostbite.

After our lunch and the humbling views of the mountain that we would attempt to climb, we returned to Confluencia for dinner. Today's high air temperature was 59 degrees Fahrenheit on a clear bluebird day. In the morning, we will break camp and continue our trek up to Base Camp at Plaza de Mulas. The Horcones Inferior River flows down from the South Face and meets the Horcones Superior River where they form a confluence, hence the name of the camp.

The route up follows the Horcones Superior River for several miles and then rises onto a lateral moraine next to the Horcones Glacier. The Playa Ancha is the very wide Horcones Riverbed covered with rock and occasionally you see or cross the small stream in this very dry canyon. The route eventually narrows as it proceeds upward from 10,000 feet at Confluencia to the base of the moraine where the trail switchbacks up and over.

When you look back for miles down the riverbed, it looks like it could have been glacier carved. The Horcones Glacier did, in the past, extend down toward Confluencia from the South Face. I wonder to myself but do not ask, "How far down the valley did the Horcones Glacier once advance? Was there another glacier once in the Playa Ancha?

Information on Aconcagua's glaciers is difficult to find. A worldwide retreat of glaciers was observed during the 20th century and glaciers in Argentina, the Andes Mountains, and in Patagonia were no exception. Large glaciers like the Upsala Glacier in Argentina are monitored now regularly from the Space Station.

Specifically, the Ventisquero Horcones Glacier appears to have melted significantly on Aconcagua's South Face. The popular climbing route on the Northeast side of Aconcagua on the large Polish Glacier is still in use and I found no information on that glacier's status.

My only other wildlife sighting on our approach was a couple of very large jackrabbits in this canyon, maybe the Patagonian cavy or hare. I've never seen a rabbit that big, and I used to raise rabbits. Well, that's another story! We stopped for lunch beneath some massive boulders about halfway through our 12+ mile hike to base camp. It is another bluebird day so far!

When we reached the switchbacks that climb over the moraine and its ridge up closer to the Plaza de Mulas base camp, a snowstorm greeted us. Well, I spoke to soon and up here at 14,000+ feet they are getting a good snow in what is their summer. We set up our tents at AMG's section of base camp trying to clean tent sites of snow.

In the dining tents we had hot tea and snacks. Most mornings over the next two weeks there will be fresh snow on the tents. If the sun shines during the day some of the new snow will melt, but basecamp has a consistent packed layer of snow this year. I have seen pictures in recent years when the summit had little snow on it.

> *Climate change in Argentina is predicted to have significant effects on the living conditions in Argentina. The climate is changing primarily with regards to precipitation patterns and temperatures. The highest increases in the precipitation (1960–2010) have occurred in the eastern parts of the country. In the northern part of the country, these variable patterns have led to a higher risk of prolonged droughts, affecting agriculture in these regions.*
>
> *Agriculture will be affected by climate change. The decrease in precipitation that has been observed in the Andes is predicted to continue to decrease, affecting hydroelectric energy even more. Glaciers are predicted to continue to recede and melt or in some areas disappear. Higher*

Aconcagua, Argentina, 2018

temperatures will cause the snow cover to melt earlier in the year, causing a rise in river flow in the spring months and a drop in summer, which is when water demand is the highest for agriculture.[32]

The first day at base camp was a rest day but Shawn and I got in two miles of hiking above base camp. This was good acclimatization to the altitude and to the cooler temps. Three or four mess tents host multiple groups of climbers as they pass through base camp, some ascending and some descending.

The meals were great, and we talked with our team members to get better acquainted with all nine: two Russian women, a married couple and a friend from Paraguay, a guy from Canada, and three of us from the U.S.

That night we prepared our gear and heard a briefing from our guides on the next day's climb of Mt. Bonete. "Bonete" means "bonnet" in English. It is about a six-mile roundtrip climb to the summit at 5,074 meters or 16,647 feet. We wore layers of climbing clothes and were required to wear our plastic climbing boots even though alpine boots like my Asolo boots could make it just fine.

I also own some blue plastic Asolo boots that I have used on mountains including Mt. Rainier and Mt. Hood. For this climb I decided to buy some newer and lighter Mammut Norwand boots and wearing them on Mt. Bonete would continue to form-fit them to my feet. From the summit of Bonete there were great views of Aconcagua and surrounding mountains in the Andes as far as you could see. I felt strong physically and I really enjoyed this climb above 5000 meters.

After returning to base camp, each of our team members had some tests run on them in the medical tent. My first medcheck recorded blood pressure at 120/80 and oxygen (O2) at 92. Those were good numbers! We were encouraged to eat as much as we could (despite experiencing a loss of appetite at altitude) and to drink lots of hot tea and water.

The next day (Thursday) we made the acclimatization trek to Camp 1 known as Camp Canada at about 5000 meters. We departed base camp at 8:00 am onto very steep, snow-covered slopes carrying food, fuel, and supplies for the higher camps on the mountain.

The Argentinian porters and guides carried much heavier loads than we did. We cached our loads, ate some lunch, and descended to base camp by 2:00 pm covering a total of about five miles. *Climb high and sleep low.*

Friday was a rest day for our bodies to adjust some more. Shawn and I again hiked a couple of miles around base camp and up to the old Plaza de Mulas and former Ranger Station, now abandoned. Frequent avalanches and its location close to the mountain required a move lower to a larger and safer base camp.

Mule drivers on horses were constantly herding the mules with heavy loads up to Plaza de Mulas and then back down to Los Puquios. We always quickly moved aside when we heard them coming! What a great work they do—mules and drivers!

We felt rested and eager this morning to climb the 4,000 meters up to Camp Canada. Once there we set up our tents, ate more food, and rested all afternoon until dinner. We slept well at 16,000 feet—the highest I've ever slept! I slept at 15,500 feet on Kilimanjaro near Barafu Hut. Our guides often repeated, "If Kilimanjaro is a walk in the park, then Aconcagua is a kick in the stomach!"

Today we moved the provisions to Camp 2 at about 18,040 feet and hiked up almost two miles. Camp 2 is named "Nido de Condores" meaning the "Nest of Condors." Several days we saw these largest flying land birds in the Western Hemisphere circling above camp.

There are two species of these raptors: the Andean condor and the California condor. Both condors are very large-winged soaring birds, the Andean condors being slightly shorter on average than the northern species, but heavier and larger in wingspan than the California condor.

Porters carried the tents, food, and equi pment to Camp 2. We helped set up the tents and later our guides brought dinner for us in our sleeping tents. Up at this elevation, Camp 2 had a latrine tent because everything had to be packed out. I will spare you the details.

Aconcagua, Argentina, 2018

We rested overnight and looked forward to another acclimatizing day to adjust to the higher altitude. There is one-third less oxygen in our bloodstream than at sea level. We mostly stayed in our tents conserving energy, reading, and hydrating.

On the restday my cardio function and heart rate, blood pressure, and oxygen measurements were great. My concern was the weather: we got about 10–12 inches of snow last night and higher on the mountain it was reported as several feet of new snow. The team member update:

1. Juan, his wife, and their friend—all three from Paraguay—decided to leave the mountain at Camp 1. The story drifted up to us that she wasn't feeling well. Shawn and I had quickly made friends with Juan who was a civil engineer and marathon runner from Asuncion and spoke excellent English. Juan recommended two books on South American Incan and Aztec history that I bought and read after I returned home: *Conquistador* by Levy and *The Last Days of the Incas* by MacQuarrie. We would miss their positive attitudes. Since the Aconcagua climb, Juan and I have kept in contact through Facebook.

2. The two Russian women—Yulia and Ylena—decided to return to base camp today from Camp 2. They were both suffering from altitude sickness above 16,000 feet and Pablo gave them shots of dexamethasone for High Altitude Cerebral Edema (HACE), which must be accompanied by descending to lower altitudes. I must emphasize that these two were strong and experienced women who had climbed the highest mountain in Europe—Mt. Elbrus (18,481 feet) in the Caucasus Range. I hiked down to base camp with Javier and the two women in about three hours.

I felt physically the best on this climb of any climb I've made. Once in base camp the reports on the radio of deep snow above Camp 2 secured my decision. The Canadian climber had turned around at Camp 3 called Colera (19,600 ft.) on the North Ridge.

He was young and fit but not prepared mentally for this team climb on Aconcagua.

Now there were two team members going for the summit—Shawn and the climber from Utah named Ryan. They were the strongest and most fit team members. I would likely have slowed them down if I had continued in the deep snow to the summit and back to Camp 3. With that much snow up high and my 68- year-old body, I only partially regret my decision.

> *The climate of Aconcagua is mainly determined by its height and location. Because of the low-pressure system coming off the Pacific Ocean, the central Argentina Andes, and particularly Aconcagua, receive humid winds from the anticyclone areas of the Pacific Ocean. This climatic phenomenon known as Zonda wind sends warm, humid winds east toward the Andes.*
>
> *Because the mountain Aconcagua is so big, the warm Zonda winds are forced upward thereby rapidly cooling the wind. By the time it reaches the upper mountain, the once warm wind is so cold that it precipitates in the form of snow. These snowstorms occur frequently throughout the entire year, not just in winter.*
>
> *We call the Aconcagua climate a "microclimate" and when very strong winds blow in off the Pacific, they often reach the highest parts of Aconcagua (above 5500 meters). Often when this happens a big mushroom-shaped cloud is formed above the summit. If this occurs you can guarantee there are dangerously strong winds and cold conditions on the summit. Climbers should avoid the summit if they see this cloud cap as trekking into it can often be fatal.*[33]

Word began to be relayed down to us that the body of a climber (about age 60) was seen at or near the summit and that he was dead. His identity could not be released until the family was notified and his body could not be retrieved from that high on the mountain. His body would have to lie there for a week or so and perhaps left there or maybe brought down to Camp 2.

There was a helicopter landing pad for the evacuation of people at Camp 2. It was the highest altitude that the helicopter could

Aconcagua, Argentina, 2018

safely reach. During our time there I saw several people evacuated by helicopter both from Camp 2 and from Base Camp.

Since I didn't have a tent at base camp, I found an empty bunk bed in one of the bunkhouses (no heat) for the night. I woke up the next morning cold due to more snow at base camp overnight. It was usually 9:00 am or after when the sun peaked over the surrounding mountains and began to warm up camp.

Until that point—from especially 6 to 9 am—my hands and feet would freeze a little unless I moved them constantly. They were not exposed so frostbite wasn't the issue. I would dress and pull on my frozen Asolo boots and go hang out in the cook tent to warm up. The base camp staff members were awesome, and they let me also charge my phone there.

I would go hiking two miles every day above base camp and get my body moving. I never risked frostbite from wet or exposed hands or feet, but the sensation of very numb fingers and toes for several hours was different.

Meeting and talking with new people always seemed to help as well. I had become acquainted with a team from Texas—Kevin, Mike, and Lisa—when we first arrived at base camp. They were doing the Aconcagua climb to raise money for water wells in places in the world that need clean drinking water. Kevin and Mike went on to the summit, but Lisa had to be evacuated by helicopter today. (That ride evidently costs a climber a very pretty penny—maybe $7–8K.) She had descended the upper mountain in her own strength but when I saw her, she looked exhausted.

The freezing conditions, low humidity, and low oxygen mean that climbers often compare Aconcagua to 8000-meter peaks in the Himalayas. Many top climbers use the mountain as a warm-up for the taller peaks and a few notable climbers have stated that the Aconcagua summit feels colder than that of Everest!

On the mountain Shawn and Ryan were still waiting at Camp 3 for the weather to get better. Hanging out at base camp I was meeting people from many countries: a couple from Amsterdam; a father and his twelve-year-old son team from Colorado; a Russian team (no English speakers); two AAI guides Bryan and Juan from

Washington state; and a team from the UK—three Brits, Mike from Limerick, Ireland, and "Banjo" the Irish fireman.

I struck up a conversation with Mike and Banjo about Irishman Tom Crean, the polar explorer from the town of Anascaul on the Dingle Peninsula. Banjo was shocked and said, "How'd you know him?" I told him about my Carrauntoohil climb and we were instant friends.

By evening I had talked to Shawn by radio, with the help of Cecilia and Emilio, my new friends in the cook tent—Shawn and Ryan with guide Pablo were going for the summit on Friday! Three days at Camp 3 was making them a little crazy. I did my daily two-mile hike and saw Yulia and Ylena leaving to hike the long trail out to the Horcones TH with Javier. They told me "Goodbye" and said that I was an inspiration to them. The weather had started to clear at base camp and bring in bluebird days!

Although the normal climb is technically easy, multiple casualties occur every year on this mountain. This is due to the large numbers of climbers who make the attempt and because many climbers underestimate the objective risks of the elevation and of cold weather, which is the real challenge on this mountain. Given the weather conditions close to the summit, cold weather injuries are very common.

After summiting on Friday, the team of three—Shawn, Ryan, and Pablo—returned to base camp about 2:45 pm on Saturday. I was glad to see them, and we ate, rested, and re-packed our gear for tomorrow's descent and hike out. Shawn and I talked about the summit day, the snow, and the top of Argentina.

"It was a tough day" he said. They saw and walked around the body of the dead climber still lying near the top. Shawn does not brag at all, and I sensed a huge feeling of accomplishment in him. I was sure proud of him! *This climb has been another great father and son experience!*

We hiked about 16 miles down today to the Horcones TH where a charter bus was waiting for us. Pablo's girlfriend is an M.D. and worked some days at base camp, where we met her. A few days before we began our hike out, she got very ill at altitude and had to be helicoptered to Mendoza. Pablo was worried and very eager to get home and be with her. He asked if we could all run with our packs the four or so miles from Confluencia to the trailhead at Horcones. We said, "For you, sure!"

It was a long trip on the bus back to Mendoza with several short stops for a person to get off the bus. Shawn and I talked about more details of summit day. I told him that I had seen pictures of the summit and summit ridge when there was temporarily no snow on top. I love weather-watching, and my family will confirm that if I'm not viewing the Weather Channel then I am using the weather app on my phone!

On Monday, Shawn I walked several miles in Mendoza and found a great steakhouse that served premium Argentina beef. We both had a big steak and a lager. Later in the day our flight would "hop" over the Andes in about one hour flying time into Santiago, Chile. The Santiago to Houston flight was an overnight one arriving on Tuesday. I took my connecting flight to Little Rock and Shawn took his to Dulles in the D.C. area. We would recall this trip and talk about it often!

Back home I wrote down these notes:

> In the tables in the back of my running journal—The Runner's Training Journal—my healthy weight range for my height is 136 to 163 lbs. The actual target weight is 150 lbs. I left the States on January 1st weighing 162 and returned on January 24th weighing 150. This weight was my weight as a high school senior.

One last trip footnote: The NH Hotel misplaced my small, black Eddie Bauer backpack in storage but, after numerous calls to Acomara, Viviana retrieved it from the hotel, mailed it to me from Mendoza, and I received all my valuables about three weeks later.

Mountains have shaped human history, and they continue to shape our family history!

11

Chicago Basin, Colorado, 2018

Pine Beetles and Forest Fires

You must go on adventures to find out where you truly belong. –Sue Fitzmaurice

The Dry Fork trail in the San Juan National Forest was one of many trails closed because of the 416 fires on June 12, 2018, in Durango, Colorado. The Durango & Silverton Narrow Gauge Railroad was shut down for more than 40 days that summer and its coal-fired tourist trains may have sparked the fires that burned 53,000 acres.

In June 2020, heavy rains washed away a 40-mile section of track along the Durango & Silverton route impacting summer operations again on one of America's most scenic train routes in southwestern Colorado. The train service was cut off for eight or more weeks after a recent storm overflowed Elk Creek, causing a large buildup of debris and subsequently washing out the rail about five miles south of Silverton.[34]

Moving Mountains

In late September 2018 Shawn and I had planned our annual Colorado 14ers trip, this time to the San Juans in Southwest Colorado. The day we arrived in Durango was the first day that the train was running again, and the summer tourist train season was almost over. I'm not sure whether the train runs all year, but I've noticed online that they do have some winter and spring dates for this scenic route through the Rockies.

We had hiked in the Weminuche Wilderness many years before, but we had never climbed 14ers there. Some people say the San Juan Range is Colorado's finest range because of its vast size covering more than 4000 square miles. The San Juans contain 13 fourteeners and six wilderness areas including the Weminuche which is the largest. These mountains are rugged and some of Colorado's most difficult 14ers are here. They get a lot of winter snow and often carry a lot of snow into August.

The Chicago Basin comprises the upper portion of the Needle Creek watershed in the Needle Mountains, a subrange of the San Juan Mountains in southwest Colorado. It lies within the Weminuche Wilderness, part of the San Juan National Forest. Needle Creek is an east-side tributary of the Animas River. The upper portion of the basin is surrounded by three fourteeners: Mount Eolus, Windom Peak, and Sunlight Peak. Columbine Pass lies to the east in the lower basin.

We loaded our backpacks onto the freight car in Durango and climbed aboard the narrow-gauge train bound for Silverton. I remember riding this train as a child on Colorado vacations with my brother and our parents. I still recall sticking my head out of the open-air passenger cars and getting a face full of coal soot or cinder. We made several trips on this train during my childhood years.

On this trip we met numerous tourists just riding the train up to Silverton for lunch in that western mining town and then back to Durango the same day. I saw two backpackers get off the train—one before we did and then one who was a photographer hiking in this area. These train stops in the wilderness feel like the frontier trains stopping at remote towns in the Wild West.

Chicago Basin, Colorado, 2018

We retrieved our backpacks from the freight car at the Needleton Whistlestop after riding the train for three hours. All the trains stop at Needleton going and coming from their turnaround in Silverton. The Needleton TH is at 8212 feet and provides access to the Chicago Basin, which is the popular approach for the standard routes up four peaks above the Basin.

Needleton is deep in the Animas River Canyon 13 miles south of Silverton and you can only reach it by rail or foot. We strapped on our packs and hiked seven miles up to the Chicago Basin in about five hours between noon and 5 pm.

The Chicago Basin is a large alpine meadow at about 11,200 feet sitting right at the base of three official 14ers—Mt. Eolus and its companion North Eolus, Sunlight Peak, and Windom Peak. These wild and rugged peaks are the most remote of Colorado's fourteeners and the highest peaks in the Needle Mountains.

The remote sanctity of these peaks is what the wilderness is all about and they awaken a sense of the sacred nature of mountains. The peaks of the San Juan are popular —- a testimony that we need wilderness now more than ever. Knowledge of wilderness—the spacious views and clarity of the air—will create the desire for the preservation of wilderness.

Climate change isn't just about melting ice caps and polar bears. The Western U.S. has seen a larger increase in average temperature in the past decade than any other part of the country. This exacerbates already existing problems such as snowpack, water scarcity, drought, forest fires, and pine beetle infestation. If you've been in the mountains lately you've seen the devastation caused by the pine beetles, which have killed an area of Colorado's forests equal to over 1.5 million football fields.

Rising temperatures and fewer below freezing winter days allow them to thrive at higher elevation. Dry, dead trees increase the likelihood of intense forest fire, forcing authorities to limit access to hiking and camping. Colorado's record-breaking wildfires of the last few decades show climate change is affecting the whole state and its resources and wildlife.

Moving Mountains

As Shawn and I picked out a campsite and set up our North Face Spectrum 23 tent, we saw one other camper nearby and a young couple camped closer to a small lake and water source. It was late September, after the summer crowds are gone, back at work and back in school, so we really enjoyed the solitude of the high country.

We had met one hunter leaving the Needleton Whistlestop who had just spent a week in bow season and killed a mountain goat. The goats loved our campsite and must have known that we were not hunters. Some people say they like the salt from human urine. They will forage through your camp if you let them.

The next morning, we took light daypacks and began our climbing adventure on these San Juan peaks. It was a tough climb on our first day, but we summited Mt. Eolus (14,083) in about four hours. Eolus is named after a Greek god of the winds, and you can hear sounds like those from an Aeolian harp—sometimes only one string or wire stretched between two points—played by the wind blowing over it.

The famous stretch of ridge called the "Catwalk" is a Class 2 feature but when it narrows there is exposure on both sides of the ridge. There is a traverse on the northeast ridge and a scramble on the upper east face that is Class 3. It was a sunny day on top with great views of peaks and Shiprock to the south.

We descended to the saddle between Eolus and North Eolus and did the easy Class 3 scramble to the top of North Eolus (14,039). We descended to our campsite by 4 pm estimating that we had hiked about five miles. We rested and then Shawn collected water in our new Hydrapack bladder and water filtration system from a small pool.

We bought it for this trip because we had heard about the drought and the possibility of scarce water. We usually carry water and boil lake or river water in the mountains. This filter worked great! We boiled some water for noodle soup, a dehydrated meal, and hot chocolate.

Day 3 in the high country was another bluebird day with the overnight temp around 30 degrees Fahrenheit and the daytime

Chicago Basin, Colorado, 2018

high approaching 70 degrees F. We hiked out of camp intent on doing a combination climb of Sunlight Peak (14,059) and Windom Peak (14,082) both one and a half miles east of Eolus.

After a couple of miles of our approach on trails, the climbing to the summit gets more difficult. A Class 3 traverse brings you to a 30-foot-high summit block and the hardest move onto the very top. This route up Sunlight's South Slopes is rated a Class 4 climb due to the exposed move on the summit block.

We descended the same route to 13,100 feet in the basin between Sunlight and Windom and then climbed up to the saddle between Windom Peak and Peak 18 (13,472). The route continues along Windom's West Ridge to the summit. Windom is easier to climb than Eolus or Sunlight and yet the combination route is rated Class 4. All three peaks require scrambling to reach their summits.

We left this morning at about 7:00 am, hiked about five miles, spent a little more time route-finding, summited two peaks, and descended to camp by 5:00 pm. This was a full day, and we were hungry for one of those Mountain House beef stroganoff dinners and some hot soup.

Shawn and I have great talks on these trips and get to really catch up on family and work. Because his former work at the Pentagon and now at a defense industry company involves constantly monitoring issues and trends, we cover a wide range of topics. In the mountains the time between dinner and crawling in the tent for a night's sleep on the ground is prime time for discussing the day's efforts and current events.

While we haven't discussed climate change in detail, as a scientist and nuclear engineer I'm sure he knows much more than me! There is a nuclear energy piece to reduce carbon emissions:

> *Nuclear power is a low-carbon source of energy. In 2018, nuclear power produced about 10% of the world's electricity. Together with the expanding renewable energy sources and fuel switching from coal to gas, higher nuclear power production contributed to the leveling of global carbon dioxide emissions at 33 gigatonnes in 2019. Clearly, nuclear power—as a dispatchable low carbon source of*

Moving Mountains

electricity—can play a key role in the transition to a clean energy future.[35]

On Friday, September 28th, we broke camp, packed our gear, and began the seven-mile hike down to the Needleton TH to wait a few hours for the train. We descended in about four hours arriving at the TH by the RR tracks at 1:00 pm. We were already reflecting on this year's adventure: at least 24 miles of trail hiking; 28 hours total hiking, roughly 12,000 feet of vertical gain, and four more 14ers (three official ones).

We have never had as our goal reaching the summit of all the 54 or 56 or 58 fourteeners in Colorado. My total Colorado 14ers is now around 50 and Shawn's total is a few less as he was unable to join me on all my Colorado trips. *Going to the mountains is in our DNA. It is my hope that my children and grandchildren will continue to hike and climb in the mountain wilderness.*

There are many people who have done the whole list and some who have done all 100 Colorado peaks over 13,800 feet. Some have done the traditional list of 54 fourteeners in less than two weeks. Some runners have run them all. A few skiers have skied all 54 in the winter which means they have also climbed all 54 in winter. A man has done that, and a woman has done that. The stats are impressive, but I won't reprint them here. One climber I remember logged over 300 miles and over 150,000 feet of vertical gain finishing all 54 peaks. There are so many records. Google it.

There are lots of mountains and mountain ranges in the world and our goal has been to climb in the U.S. as well as other countries. I have hiked and climbed on all seven continents and so I still have a goal of returning to the Ellsworth Mountains of Western Antarctica. There is a quote from Patagonia founder Yvon Chouinard, alpinist and environmentalist, I particularly like: *how you climb a mountain is more important than reaching the top.*

As I've noted numerous times, the weather in the Rocky Mountains of Colorado is often spectacular in the month of September. The most important thing in the weather is knowing that mountain conditions can change quickly, and the climber must pay attention to even small changes. Many of the Colorado 14ers

Chicago Basin, Colorado, 2018

do not necessarily have a recorded first ascent (FA). I did not make an exhaustive search for them.

John Muir was an influential Scottish American naturalist, explorer, writer, and an advocate for the preservation of wilderness in the United States. Perhaps a John Muir quote can wrap up this chapter: "The clearest way into the universe is through a forest wilderness."

12

Kosciuszko, Australia, 2019

Drought, Forest Fires, and Coral Bleaching

The journey of a thousand miles begins with a single step. —Lao Tzu

It is a long way to Australia from Arkansas! June 2019 brought another International Economic Development Conference opportunity for Lou Ann's work, and we immediately agreed that if I would pay for my trip, we would travel to Australia. Her two-day conference was in Sydney and there was so much to see in that great city.

I could hike up Australia's highest mountain and after her conference and my trip to Thredbo, we could fly to northern Australia and the city of Cairns for sightseeing including going out to the Great Barrier Reef. Always explore!

After returning from Australia, we heard daily reports on the forest fires that were burning extensively and threatening and killing wildlife including wallabies, kangaroos, and koalas plus bird species and amphibians. Preceding the fires, Australia had been experiencing drought and that would become *observable* on our trip, especially as I drove out to the Snowy Mountain Range.

Author Jeffrey Bennett writes, "Drier and hotter conditions in turn lead to greater danger from wildfires, and significant increases in wildfires already have been *observed* in many places around the world, including the American West, Alaska, Canada, Australia, and Russia."

> *The 2019–2020 Australian bushfire season, colloquially known as Black Summer, was a period of unusually intense bushfires in many parts of Australia. The 2019 warning for Queensland and Northern Australia was due to exceptional dry conditions and a lack of soil moisture combined with early fires. In late 2019 smoke blanketed streets in Sydney, New South Wales.*
>
> *From September 2019 to March 2020, fires heavily impacted various regions of the state of New South Wales and in eastern and northeastern Victoria. While lightning strikes, accidents, and possible arson may have started the fires, they were likely enhanced and sped up by drought and global warming.*
>
> *Australia is one of the most fire-prone countries on earth, and bushfires form part of the natural cycle of its landscapes. However, factors such as climate trends, weather patterns, and vegetation management by humans can all contribute to the intensity of bushfire seasons. The burned area from June 2019 to May 2020 was approximately 46 million acres with a cost exceeding $103 billion. 3500 homes and 5800 outbuildings were destroyed by the bushfires.*[36]

We flew from the DFW airport on June 13, 2019, and arrived in Sydney on June 15, 2019, after crossing the International Dateline and losing a day. Well, we will get it back on our return flight! Taking the train to our city center hotel, we were eager to check-in and get in some sightseeing on foot that Saturday afternoon!

On Sunday morning we walked four or five miles along the Circular Quay on Australia's east coast taking in the famous Sydney

Kosciuszko, Australia, 2019

Opera House and the Sydney Harbour Bridge. There were many shops and restaurants and outdoor cafes in this area, supported by Sydney's 5.3 million people and throngs of tourists.

On Monday, Lou Ann's Conference was set to begin, and I rented a car from Avis Rental and began my long drive to the mountains. There was plenty of traffic as I left Sydney, but fortunately more cars and trucks were entering the city while I was exiting headed for open spaces. Along this mostly southward route, you see lots of kangaroos roaming the vast countryside and many dead ones along the highway.

Due to the extremely dry conditions and widespread drought, the animals would graze along the road and even find dew or moisture on the road's edges where they would be struck by vehicles. I never hit a kangaroo. I crossed over the Great Dividing Range and turned a little westward toward the town of Jindabyne and on to Thredbo which is a snow skiing village in the Kosciuszko National Park in the Main Range of the Snowy Mountains.

Mt. Kosciuszko is mainland Australia's highest mountain at 7310 feet asl. The peak is usually approached from Thredbo, taking three hours for a round trip. This easy walk starts from the top of the Thredbo Kosciuszko Express chairlift which operates year-round and will cost you an exorbitantly priced lift ticket like most prices we paid in Australia.

The walking path is popular in summer, but I saw scarcely a soul on my winter hike. The "trail" was dirt and snow but primarily a steel mesh walkway built to protect the native vegetation and prevent erosion. The walk to the summit is the easiest of all the Seven Summits, made popular by that mountain challenge first completed in 1985 by Dick Bass. I was up and down in less than three hours and got a room at the Thredbo Alpine Hotel. Aboriginal people probably made the first ascent of Kosciuszko!

Kosciuszko is the highest mountain on Australia's mainland but purists and those climbers after Dick Bass argue that it is not the highest mountain in the Australian continent which includes islands and New Guinea. Puncak Jaya is in the Indonesian province of Papua on the island of Papua, which lies on the Australian continental shelf at 16,024 feet asl. Carstensz Pyramid, as Puncak Jaya is also known, is a mountaineering expedition.

One of my mountaineering heroes, climber and photographer Galen Rowell, used this expression, "Let's drink some coffee and get after it!" The next morning, I walked a couple of miles in Thredbo and loaded up on morning coffee before starting the long drive back to Sydney. This summit was important to me as one of the Seven Summits, but it was a lot of driving for an easy 7+ mile walk.

I guess it made me desire to visit Papua, New Guinea, and attempt Carstensz Pyramid. The drive back was a relaxing drive in the country to the outskirts of Sydney where the traffic was fast and furious as I searched for the city center exit on my GPS.

I joined Lou Ann for an evening cruise in Sydney Harbour with folks from her economic development crowd. The nighttime views were spectacular and so were the food items served on board the boat. Later we did some walking in the Rocks District and then went to bed to rest for our flight the next day to Cairns and the Cairns Beach Resort.

Cairns is a city of 150,000+ located in Queensland on the east coast of Far North Queensland. It was founded in 1876 and named after the Governor of Queensland. It was formed to serve miners heading for the Hodgkinson River goldmine but declined when an easier route was discovered from Port Douglas. It later developed as a railhead and major port for exports.

> *In addition to the drier weather, droughts, and higher temperatures that Australia has been experiencing increasingly for the past two decades, predictions of scientists assert that global warming will negatively impact the Australian continent's environment, economy, and communities.*
>
> *It is vulnerable as a country because of its extensive arid and semi-arid areas, an already warm climate, rainfall variability, and existing pressures on water supply. Australia's population is highly concentrated in coastal areas, and its important tourism industry depends on the health of the Great Barrier Reef and other fragile ecosystems.*[37]

Kosciuszko, Australia, 2019

The Holloway Beachfront at our resort on the Captain Cook Highway in Cairns was lovely and we walked numerous times a day during our four-day stay. The population was small here despite being at the edge of the city. There was a small beach shack restaurant serving breakfast, lunch, and supper right on the beach where we could eat, relax, and enjoy the coastal views.

Toward town about one mile there was a grocery market where we could buy food and supplies. The Cairns Beach Hotel where we were staying right across the street from the beach had a beautiful swimming pool and cabana. *We might not go home!*

The married couple who owned and managed the hotel had plenty of recommendations for excursions and outdoor adventures for us. From zip lines, water sports and scuba diving, whitewater rafting, kitesurfing, and skydiving to exploring the rainforest, the options seemed endless. We listened and we read lots of brochures relaxing beside the pool. Finally, we focused on taking the Quicksilver cruise out the Great Barrier Reef and doing a Wilderness EcoSafari in the rainforest.

Kevin Simpson, owner and operator of Wilderness EcoSafaris, described as "Crocodile Dundee, Steve Irwin, and Encyclopedia Botanica combined," was our guide on one of the best excursions we've ever taken. We were his only clients on Friday, June 21st, and we thoroughly enjoyed his humor, knowledge, and personal approach to nature.

His company had exclusive restricted access to rainforest areas and comfortable 4WD vehicles for exploring on dirt roads and mountainous Cairns Highlands outback. He was part actor-entertainer and part forest ranger, here in his own words:

"You will experience World Heritage Wet Tropics Rainforest from a unique perspective with our 'ecosensitive' 4 Wheel Drive Safari. We take you away from the mass market tourist areas, into a spectacular and ancient wilderness wonderland offering the most bio-diverse ecosystem in the world. You will venture with us into 'restricted access' areas where you will find no other crowds or vehicles, no sealed roads, just the genuine 'Land of the Giants', as you discover the largest and most ancient rainforest trees and plants in the world (not seen by other tours), some of which date back over 300 million years."

It sounds like some hype or marketing "mojo," but we saw all this and more. Kevin drove us first through the rainforest pointing out unique trees and plants and we walked a few forest paths to see creeks and waterfalls before reaching a locked gate into Barron Gorge National Park. Kevin had a key, of course, and we entered by jeep into the promised restricted area. We stopped at Lake Morris in the Lamb Range with spectacular scenery all around and had a snack and morning tea provided by Kevin.

"Queenland's Wet Tropics World Heritage Area protects and conserves our remaining tropical rainforest for the present and future generations. We all share a responsibility for their care."—A signpost at Lake Morris.

From there we drove through the rainforest to Davies Creek Falls in pristine wilderness with flowing creeks and thick canopy. We lingered in this area and Kevin produced an alfresco Ploughmans lunch in this tranquil setting. Lou Ann and I both thought the lunch was delicious.

Kevin was always alert to animal tracks and wildlife spotting and we observed most of the wildlife we anticipated: platypus, kangaroos, reptiles including rainforest dragons, turtles, cassowaries, and other unique birds. We saw giant and ancient Mangrove, Fig, and Eucalyptus trees and the Banyan tree whose roots grow up and hang down, and, in size, reminded me of the giant redwoods in California.

In the afternoon we drove on into the wilderness to rock formations and cliffs at Mareeba. Kevin encouraged us to swim in the cold creeks there (Stoney Creek I think) and so I stripped down to my Ex-officio underwear and took a swim. It was cold! This was the only place that we saw any people and there were just one or two at this spot. Kevin never meets a stranger and so he easily strikes up conversations with new people.

As promised when we left the Park, Kevin took us to visit a local creamery, and we enjoyed excellent ice cream on the edge of

Kosciuszko, Australia, 2019

the Outback. Lou Ann and I are business owners and for ten years we owned a natural market like Whole Foods or Fresh Market.

We had a great visit with the couple who owned the creamery, and we will never forget their story of the Australian government pushing up the national minimum wage to $20/hour. It had severely hurt their ability to employee people in their business and threatened the future of their business and other small businesses, the backbone of a market economy.

Driving back into Cairns in the late afternoon at the end of our nine-hour excursion we saw hundreds of kangaroos grazing in the open fields. Australia has a lot of them somewhat like our deer populations in some urban areas of the country. *This was a special day! Thanks Kevin!*

Lou Ann and I finished up this awesome day with dinner at the beach shack and a walk on Holloway Beach. There was a lot to talk about and to absorb from this day. We saw some amazing nature unlike anything we had ever seen before. We learned new things about the earth and the rainforest. We renewed our commitment to preserve this for our children and future generations. *We feel extremely fortunate to have these travel experiences!*

The next morning after a breakfast of pastries and coffee in our hotel room, we walked several miles on the beach earlier than usual. We rode up the coast for about an hour on the Quicksilver coach to Port Douglas where we departed by boat for the Great Barrier Reef and Great Barrier Reef Marine Park.

I have always wanted to see the GBR, but I had no idea that it extends 2300km (1426 miles) along the northeastern coast of the Coral Sea. It is not one reef but a network of about 2900 individual reefs stretching from Cape York in the north to Bundaberg in southern Queensland. The Great Barrier Reef is a World Heritage Site.

Port Douglas looked like a charming seaside village and as we sailed out of the port in the Quicksilver Wavepiercer we could see the old wharf and a building that remains from the early 20th century. Port Douglas back then was a thriving port serving the goldfields and Tablelands inland and the sugar industry in the Mossman region.

Out in the bay the Mossman and Daintree Rivers empty—the Mossman River is known for its beautiful Gorge and the Daintree River is known for its diversity of wildlife. Both rivers are fringed by a spectacular rainforest.

Captain James Cook sailed these waters in the vessel "Endeavour" on June 11, 1770, and struck a reef 30 miles north of Snapper Island. We passed the Undine Cay which is an example of a coral cay in its early development and signposts the inner edge of the Great Barrier Reef.

After passing over the GBR on Australia's continental shelf, the small group of ribbon reefs, called the Agincourt Reefs, is on the very outer edge of the GBR. Two kilometers beyond this, identified by a line of breakers, the sea floor drops away to a water depth of more than 500 meters.

The clear, clean water from the deep ocean washing over these outer barriers helps promote the prolific growth of marine life for which the GBR is famous. *Climate change is one of the greatest threats to the future of the Great Barrier Reef.* That statement was one for which I was totally unprepared. I had no idea. The damage from rising sea temperatures and ocean acidification was already happening. Seeing the GBR and its fragile marine environment in person was eye-opening!

Engaging the dramatic beauty of the Great Barrier Reef is an extraordinary experience and it's a day you will remember because it engages all your senses, lifts your spirit, and takes you into another realm. Sharing the wonder of the world's largest living entity and seeing nature in all its underwater majesty, was a true exploration. The Great Barrier Reef vibrates with life in a coral garden filled with marine life.

Quicksilver not only shuttles tourists and scientists the 45 to 50 minutes from Port Douglas to the GBR but they also operate a stable reef platform for tourist activities and for research. Lou Ann and I first chose to view the coral reefs in a semi-submersible vessel for a closer look at the different types of coral—Staghorn,

Kosciuszko, Australia, 2019

Brain, Mushroom, and Plate Coral—and many species of fish. On the ship and on the reef platform we listened to marine biologists explaining the coral reef and the threats to it.

> As our oceans become warmer and more acidic, the Reef has suffered. Coral reefs depend on the colorful algae that live throughout their nooks and crannies to survive. But when it becomes stressed by warmer and more acidic waters, coral expels these algae and turns white—a phenomenon known as "coral bleaching." Bleaching weakens the coral, and it may begin to starve. Unless the coral has a chance to recover and the algae can return, it can die.
>
> This can happen on a vast scale, in what's called a "mass bleaching event." In 2016, the Reef experienced a mass bleaching event that scientists say was made 175 times more likely by climate change. Another bleaching event hit the Reef in 2017, bleaching even more corals.[38]

Since Lou Ann and I were not scuba divers, we opted next for a few hours of snorkeling all around the reef platform. The experience was awesome in so many ways—observing the bleached coral, swimming around so many colorful fish, and spending so much time in the water.

We also viewed the coral and the fish from the platform's underwater observatory which had pictures on the walls to identify fish and other marine life. We had morning and afternoon tea plus a wonderful hot and cold tropical buffet lunch with prawns.

In late afternoon the Wavepiercer returned us to Port Douglas in about an hour from the Agincourt Reef Platform. This was a trip highlight and yet, an eye-opening experience for me. We couldn't have had a more meaningful and enjoyable excursion than this ten-hour adventure. We talked about this amazing outing all evening and during our walk on the beach the next morning. We enjoyed the brilliant weather.

We left Australia for our flight to DFW on June 23rd and we arrived home on June 23rd. *Now that is expeditious travel! And we found ourselves back in Arkansas passing through familiar towns but with a fresh vision.*

13

Antarctic Peninsula, Antarctica, 2023

Rising Ocean Temperatures

Antarctica is a barometer of global climatic health.—Tom Griffiths

During the Heroic Age of Polar Exploration, 1897–1922, the Antarctic region became the focus of international efforts that resulted in intensive scientific and geographical exploration by 17 major Antarctic expeditions launched from ten countries. These were recorded in the books I read growing up and the heroes' names I heard—Shackleton, Crean, Amundsen, Hanssen, Scott, Hayward.

Becoming an Eagle Scout, an Explorer, and a select member of the Order of the Arrow in the 1960s, the great outdoors— the forest and fields, the mountains and glaciers, the rivers and lakes— always feel like home. I pledged an oath in those tender years to protect our natural resources for future generations.

Moving Mountains

Commenting on my Antarctica trip in 2023, well-meaning people said things like: "a trip of a lifetime" and "living your best life." My reply: "It was more than that." This is my Father's world, and I travel this Earth that God created to visit the countries and continents because I want to be filled with an intense interest in everything. I have been to six continents at least two times each. Antarctica is my seventh continent, and I plan to go there again despite the distance and the cost. Never stop exploring!

Our 140 travel companions and the 150+ expedition staff and crew explored together because "community" is relational. We had no internet for nine days and dined at tables together for 26 meals. We met and made many new friends, and these people are the best part of life, gifts from God on our journey. This story is about the white continent, its wildlife, and the natural resources we need to conserve and protect. I am now a member of the Shackleton Club. *Good explorers respect their environment; great ones protect it. (Quark)*

Rapidly melting Antarctic ice is causing a dramatic slowdown in deep ocean currents and could have a disastrous effect on climate. The currents carry vital heat, oxygen, carbon, and nutrients around the globe. Crossing the Drake Passage and the Southern Oceans in early March 2023, we were introduced in person to the Antarctic Convergence, the Antarctic Circumpolar Current (ACC) that flows clockwise around the entire continent, and the Antarctic Circle Boundary.

The ACC is the strongest current system in the world oceans and the only ocean current linking all major oceans: the Atlantic, Indian, and the Pacific Oceans. The Antarctic Circle divides Antarctica from the Southern Temperate Zone to the north. Yes, the Southern Ocean rocks and rolls! Imagine the early polar explorers crossing it in wooden ships. *It is a pure realm of water and wind, an extensive belt of the Earth's surface, where these two fluids encounter and animate one another without interruption.* (Griffiths)

I am not a cruise ship fan and have little experience crossing the oceans except by aircraft. On one occasion my wife, our

Antarctic Peninsula, Antarctica, 2023

daughter Savanah, and I took a large cruise ship from Mobile, AL, to Cancun, Mexico. One and done!

I am fascinated by the ocean. We have kayaked and snorkeled in Hawaii, Puerto Rico, and the Dominican Republic; observed calving glaciers and whales in the Gulf of Alaska and the Kenai Fjords; identified the North Atlantic Current off the coast of Ireland as well as the Azores Current and Gulf Stream near the Azores Islands.

We spent a summer on the Costa del Sol of Spain teaching college courses while drinking cafe con leche beside the Mediterranean Sea. I've seen the Atlantic Ocean from West Africa and the Indian Ocean from East Africa. World Geography was my favorite course as an undergraduate at West Texas A&M.

I am a mountaineer, not a sailor. I am far from resembling this sentiment of Herman Melville, but his energy I like: *Give me this glorious ocean life, this salt-sea life.* I identify more with this feeling: "The wonderfully clear days, the sight of a new rock beyond the snow rise, the tremendous feeling of freedom among the mountains and glaciers, the close comradeship which develops in isolated groups from shared experience and the growth of mutual confidence: these are lasting memories."—Of Ice and Men, V. E. Fuchs. Both quotes embed the idea that we can be Always Exploring.

Antarctica is astonishing, drawing me inland to climb the Ellsworth Mountains or further south at Mt. Erebus on Ross Island. The Antarctic Continent is vast, silent, and white, yes, highlighted with colorful lichen of red, gray, green, and brown. The views are stunning and almost impossible to grasp, words are not helpful. Asked what word or phrase comes to mind, I told our Expedition Leader, Shane, that I kept seeing "white satin" as I gazed across the vast continent and singing the words to "Calypso" by John Denver.

The White Wilderness seems at first silent and frozen, then its stillness is broken by the sound of ocean waves crashing onto polished rock shorelines. The glaciers calve and slide into the sea as we kayak nearby. The milky water floating with pack ice is just above freezing here in late summer/early fall. It will soon freeze back over as the days shorten by 8–9 minutes every day now. If you fall into the sea here, you will become one with the sea.

The Earth's network of ocean currents is partly driven by the downward movement of cold, dense saltwater toward the seabed near Antarctica. But as fresh water from the ice cap melts, sea water becomes less salty and less dense, and the downward movement slows. These deep ocean currents, or "overturnings," in the northern and southern hemispheres have been relatively stable for thousands of years, scientists say, but they are now being disrupted by the warming climate.

Sailing on the ship Ocean Diamond from the southernmost city of Ushuaia in early March, we entered the Beagle Channel after sunset. Before it was totally dark, we spotted humpback whales and sea lions in this 150-mile strait in the Tierra del Fuego Archipelago. This is the extreme southern tip of South America between Chile and Argentina.

The Beagle Channel, the Straits of Magellan to the north, and the Drake Passage to the south are the three navigable passages around South America between the Pacific and the Atlantic Oceans. We were sailing for the open-ocean Drake Passage, named after the 16th Century English explorer, Sir Francis Drake.

The Drake Passage is considered one of the most treacherous voyages for ships to make. Currents at its latitude meet no resistance from any landmass, and waves top twelve meters (40 feet), hence its reputation as "the most powerful convergence of seas." As the narrowest passage around and to Antarctica, its existence and shape strongly influence the circulation of water around Antarctica and the global circulation, as well as the global climate.

It took us two and a half days and nights to cross the Drake to our first landing on the Antarctica Peninsula at Port Charcot. We experienced some 12-foot waves at times, but our crossing was not as bad as we had anticipated. While some passengers did get seasick, Lou Ann and I experienced no seasickness.

We walked every day on decks 5 and 6, which required ten laps to cover one mile. Some days we faced snowstorms and strong winds. We watched so many Albatrosses flying around the ship as

Antarctic Peninsula, Antarctica, 2023

we sailed and learned that the birds could fly for weeks or months, never landing.

Though very different from my longer mountaineering expeditions in Wyoming, Tanzania, and Argentina, this 14-day adventure truly had an expedition feel to it. From our sea kayaking guides, Zodiac drivers and guides, wildlife scientists and photographers, polar exploration historians, glacial geologists to the Ocean Diamond's crew and staff, we were provisioned to encounter the 7th Continent. What a great team!

In Tom Griffiths' book, Slicing the Silence: Voyaging to Antarctica, he describes the White Continent:

> *Antarctica has an aura. I don't just mean its mystery, its magical otherworldliness, its implacable grandeur, and its capacity to haunt all who have visited it. I also mean that this land over the South Pole, which is covered by a single mineral, actually emanates ice, water, and air well beyond its geographical boundaries. It took people a long time to realize that there was not just land down there but a continent, that it was high and dry and covered thickly in ice, that it was very, very cold, much colder than the Arctic. And that it constantly affects the climate of the rest of the world.*[39]

The scientific body that advises the United Nations General Assembly on rising temperatures has just released a new report. It's an important summary of six key pieces of research completed over the past years. (BBC) Number Four—Our actions now will resonate for thousands of years—really caught my attention:

> *It's amazing to think that the decisions we make around the world over the next seven years will echo down the centuries. The report warns that with sustained warming of between 2 and 3C, the Greenland and West Antarctic ice sheets will be lost "almost completely and irreversibly over multiple millenia. Many other thresholds will be crossed at low levels of heating, impacting things like the world's glaciers.*[40]

Moving Mountains

On our 2023 voyage to the Antarctica Peninsula, we were primarily visiting the islands off Graham Land, hundreds of miles north of the Ross Ice Shelf which has been in the news regarding melting ice sheets. From our Zodiacs and Sea Kayaks we did see some retreating glaciers and melting pack ice late in the summer season. Our guides, who had traveled there for many years, did point out some climate changes affecting the wildlife and the glacial environment.

Glaciologist Claude Lorius led 22 expeditions to Greenland and Antarctica during his lifetime and his research helped prove that humans were contributing to global warming. The more polar expeditions he led to the continent, the more he became fascinated with Antarctica's mysteries.

He decided to study ice cores—samples drilled out of the ice— creating 160,000 years' worth of glacial records. *They showed that, for long periods of time, levels of carbon dioxide varied slightly but after the Industrial Revolution concentrations of the greenhouse gas had rocketed as temperatures rose. His research left no doubt that global warming was due to man-made pollution.*

Dr. James B. McClintock, one of the world's foremost experts on Antarctica, has created a presentation entitled, "From Penguins to Plankton— the Dramatic Impacts of Climate Change on the Antarctic Peninsula." Western Antarctica is experiencing record warming temperatures that are causing glacial recessions and ice shelf break-ups. Annual sea ice is rapidly receding with warming seas. Sea ice-dependent marine animals are being impacted by these dramatic changes to their habitats.

Our Antarctic Operator, Quark Expeditions, provided us with a post-voyage wildlife list of birds and marine animals that we observed over our nine-day adventure. The list included four species of penquins, four species of albatross, ten species of petrels and shearwaters, three species of cormorants, one snowy sheathbill, three species of skuas, and five species of gulls and terns.

The Marine Mammal list was comprised of Antarctic Minke Whale, Humpback Whale, South American sea lions and fur seals,

Antarctic Peninsula, Antarctica, 2023

Antarctic fur seal, Leopard seal, Southern Elephant seal, Weddell seals, and Crabeater seals.

We returned home very thankful for this remarkable journey, for the friends we made, and for the knowledge we gained. All over the world we see impacts of changing climates from flooding and torrential rainfall, drought, glacial melting, ocean warming and acidification, among other effects. Antarctica is a long way from home and distant from most populated places, but its climate has always determined weather and meteorological conditions in those distant places.

We traveled nearly 6000 miles by air and 1644 nautical miles to explore the seventh continent and to experience first-hand the unique role that Antarctica plays in the global climate. I suppose we could have read these facts in a book but seeing them with our own eyes makes them real and personal. We are not only new members of the Shackleton Club, but we are also now Ambassadors for the Antarctica Wilderness and the Southern Oceans.

Good explorers respect their environment, Great ones protect it.

> *Be responsible for fish in the sea and birds in the air, for every living thing that moves on the face of the Earth.*
> *—Genesis 1:28*

14

Mount Rainier, Washington, 2007

More than Melting Glaciers

Heading into the mountains with your life strapped to your back is the ultimate expression of freedom. –Jonathon Dorn

---

This chapter is out of chronological order, but I decided to return to the time frame when I started this story and include my climb of Mt. Rainier in June 2007. I've shared adventures from all seven continents and numerous climbs in the Colorado Rockies plus Wyoming in this book. Other than a short excerpt from my Mount Hood climb, my trips to climb in the Pacific Northwest on glaciated mountains including Mt. Shasta and Mt. Rainier are an important part of the larger story. (My Mt. Shasta climb comprises a chapter in my book, *Retreat Upward: A Mountain Pathway for the Soul*, Wifp & Stock, 2024.)

Mount Rainier, also known as Tahoma, is a large active stratovolcano in the Cascade Range of the Pacific Northwest in the United States. With a summit elevation of 14,411 ft, it is the highest mountain in the state of Washington and the tallest in the Cascade Volcanic Arc. The first ascent (FA) was recorded in 1870.

Even the easiest route is a rock, ice and snow climb via Disappointment Cleaver, a rock ridge between glaciers. With 26 major glaciers and 36 square miles of permanent snowfields, Mount Rainier is the most heavily glaciated peak in the lower 48 states. I am one of the 439,460 people who climbed Rainier between 1950 and 2018, .0012% of USA population. Approximately 84 people died in mountaineering accidents in that same time.

Due to its high probability of an eruption in the near future and proximity to a major urban area, Mount Rainier is considered one of the most dangerous volcanoes in the world, and it is on the Decade Volcano list, sixteen volcanoes worldwide capable of significant destruction. The large amount of glacial ice means that Rainier could produce massive lahars (violent mud or debris flow) that could threaten the entire Puyallup River valley, as well as other river valleys.

(Why I love the Mountain West: A total of 477 mountain summits in the U.S. meet both criteria for the definition of "major summit" used here: at least 3000 meters (9843 ft) of topographic elevation and at least 500 meters (1640 ft) of topographic prominence. Of these 477 summits, 117 are in Colorado, 67 in Alaska, 51 in California, 43 in Wyoming, 42 in Montana, 40 in Utah, 38 in Nevada, 36 in Idaho, 26 in New Mexico, five in Arizona, five in Oregon, four in Washington, and three in Hawai'i.)

After climbing Mt. Whitney's East Face in Summer 2006 with an Alpine Ascents International (AAI) team, Kent Barnard and I decided the time was right for a Mount Rainier climb in June 2007. We signed on with Rainier Mountaineering, Inc (RMI) for their three-day guided climb via Camp Muir and the Ingraham Glacier or Disappointment Cleaver (DC) route to the mountain's summit.

Guided climbing leads to a guided mentality (Viesturs), but we were still learning mountaineering skills. Leading and soloing would come later. *Just because you love the mountains doesn't mean the mountains love you.—Lou Whittaker*

Mount Rainier, Washington, 2007

The DC route is the undisputed classic climb of Rainier and the best introduction to alpine mountaineering in the lower 48 states. Author and climber Ed Viesturs has reached Rainier's summit over 200 times while the current known record holder, Brent Okita, has 520 times reached the top of Mt. Rainier! Ed went on to be the first American and the fifth person to climb all fourteen of the world's 8000-meter peaks without supplemental oxygen.

In 1969 RMI was formed by Lou Whittaker who partnered with Jerry Lynch to establish a guide service dedicated to teaching and leading climbers. Twin brothers Jim and Lou Whittaker of Seattle grew up climbing in the Cascade Mountains. Jim and Lou were both invited in 1963 to join the American Expedition to Mt. Everest. Lou declined to go, but Jim Whittaker went on the expedition and became the first American to reach the summit of Everest. In 1975, Lou's son Peter Whittaker started guiding climbers on Mount Rainier, and we had the pleasure of meeting him and visiting some in 2007.

I took a flight to Portland, Oregon, on May 31, 2007, and my climbing partner Kent Barnard from Bakersfield met me at the airport for the drive to Rainier. We were planning to climb Mt. Rainier first and then Mt. Hood to accommodate my return flight out of Portland to DFW. Arriving at the RMI and Whittaker Mountaineering Guide Shop in Ashford by afternoon, we began preparations and heard instructions for our climb which included a one-day mountaineering school on the next day.

On the morning of June 1st, we rode the RMI shuttle bus with other climbers up the mountain road to Paradise Lodge for the day of training in mountaineering skills on snow and ice. We hiked and trekked all over the snow slopes above Paradise with our guides Tino and Joshua. We practiced climbing safety and climbing techniques: using an ice axe, attaching crampons, installing your climbing harness, and covered skills including roped glacial travel, running belay, self-arrest, and team arrest.

We learned techniques like the traverse step, duckwalk, cross-over step, plunge step, and switchback turns on snow. We worked (and acclimated) steadily from 9:30 to 3:30 pm with few rest breaks. We learned by ascending slopes using pressure-breathing and rest-stepping skills. We descended the snow on our backs and practiced stopping a fall with self-arrest by ice axe. The hot meal and warm bed felt good that night!

Day Two we began our climb, first taking the shuttle bus from the RMI base camp to the Paradise parking lot near the Lodge. We left the Paradise staging area around 10:00 am and hiked up to Camp Muir in plastic mountaineering boots (Asolo for me). We broke up this ascent by doing five 1000-foot sections covering about 5.2 miles and 4600 feet of vertical gain. The bunkhouse at Camp Muir was basic and crowded with about 20 climbers in several RMI groups. Kent and I were teamed with senior guide Sean and trainer Tino and each team had a 1:3 ratio of lead guide to climbers. Soup for supper!

We were awakened at 12:30 am on June 3rd to get our clothing and gear on, have some tea and light breakfast, and depart Camp Muir by 2:00 on the route up and across a near level traverse of the upper Cowlitz Glacier. We then ascended the slope through the central gap in Cathedral Rocks and climbed a long scree slope to the ridge crest on to the Ingraham Glacier and DC route. We had strapped on our crampons and climbing harnesses, wearing helmets and carrying ice axes for the roped travel of 4300 vertical feet and 2.8 miles to the summit.

Our first two days on the mountain were clear, bluebird days. Serious climbers always pay attention to the weather, but we were new to climbing the Cascade Mountains, which are not only glaciated but also situated very close to the Pacific Ocean. I should have read the material on Mt. Rainier more carefully: *The formation of a cloudcap can create hazardous visibility problems. Cloudcaps can form rapidly and envelope the upper mountain region, making progress up or down difficult. The appearance of lenticular clouds is a signal of a possible cloudcap and should not be ignored.*

We crossed 15–16 crevasses, several snowbridges, and steep sections on our way to the top. One of the roped teams below us had a climber fall into a crevasse and our guides assisted with his rescue which delayed us for maybe an hour. We reached the

Mount Rainier, Washington, 2007

summit crater rim and then the Columbia Crest, which is the true summit around 8:00 am The formation of a cloudcap had formed on the top of the mountain where it was very windy with whiteout conditions. The climb was more difficult due to the weather, and we enjoyed no view from the top. We didn't stay long!

My new plastic Asolo boots chewed up my toes, heels, and shinbones especially on the long descent back down to Paradise. I estimated that we covered about 11 miles on day three, all on snow. This was an aerobic workout for sure with periods of anaerobic effort and step-breathe movement. Kent had exercised and prepared better than me for this climb. He looked over at me once on the ascent and remarked that I had evidently not been running like I normally did to train for the mountains. Lesson learned.

> *Here on the mountain the air is clear, and your mind is clear; as you drop down into Narnia, the air will thicken, take great care that it does not confuse your mind.*
> —C. S. Lewis

Despite deteriorating weather and physical fatigue, I typically find mental clarity on the summit of significant mountains. Altitude and conditions like cerebral edema can certainly cloud your thinking and fellow climbers must keep alert for those symptoms in themselves and others. Ordinarily there exists more confusion where there are more distractions in our daily work lives. On the mountain, the air and the mind are clear. Climbing mountains has taught me to pay more attention to weather and to climate.

Scientists are studying climate change in many ways to help guide the future of Mount Rainier National Park. Climate change science at MRNP is more than just the study of melting glaciers. How will the changing climate affect rivers and subalpine meadows? Wildlife and visitor access? The Wilderness Act of 1964 defines wilderness as an area where Earth and its community of life are "untrammeled by man," but the Act also requires that managers preserve and protect wilderness in its natural condition.

As stewards of the earth, all humans have this responsibility to preserve and protect wilderness and the wild places of natural beauty. More than 97% of MRNP is legally designated as wilderness, which includes glaciers, forests, meadows, lakes, and other wetlands. While enduring impacts from climate change, wilderness is also recognized as a strong defense against it. A forest keeps carbon from becoming available as a "greenhouse" gas that raises the Earth's temperature.

In November 2023 the Seattle Times published an article summarizing "What the National Climate Report Says about the Northwest" including these observations and comments:

> *Like the rest of the world, the Northwest is at risk. Washinton, Oregon, and Idaho are home to some 14 million people and 43 Native American tribes. The region is already experiencing climate change and more will come in the decades to ahead. But the region is not without influence or options. States like Washington are scrambling to cut greenhouse gas emissions as quickly and painlessly as possible, with mixed degrees of success and local opposition. Others resist the change or even lay groundwork for the continued reliance on the fossil fuel that have brought us to this point.*[41]

Here are five things to know about the Northwest from the latest National Climate Assessment:

1. Rising temperatures—Extremely hot days are becoming more common.

2. The consequences of climate change won't be spread evenly—Extreme heat and wildfire smoke impact some populations more than others.

3. There will be winners and losers in the plant and animal kingdoms—These impacts affect food chains, soil and landscapes, and water resources.

4. Food sources and regional economies are at risk—Snow seasons in portions of the Cascades are projected to halve, impacting winter sports, hiking and camping, fishing and boating.

Mount Rainier, Washington, 2007

5. Cities, roads, and electrical grid must adapt—local and regional citizens can lead the way and create a legacy, adapting to climate change.

I love mountain wilderness, and I love the MRNP statement that "Wilderness is recognized as a strong defense against climate change." Mount Rainier and the surrounding wilderness is one of the most beautiful places I've hiked and climbed. It is covered with nearly 30 square miles of glaciers and icy patches—more than Mount Hood, Crater Lake, and all other volcanic mountains combined, from British Columbia to Northern California.

And climate change is taking a toll on Mount Rainier's glaciers, according to a study published in June 2023. It found a 42% reduction in glacier area from 1896 to 2021, and officially removed Stevens Glacier from the park's inventory. This kind of glacial reduction has been accelerating during the last twenty years.

Returning to my Mount Rainier climb of seventeen years ago now and concluding this book on Moving Mountains with that great outdoor experience and the current facts on climate change, serve to emphasize the role of wilderness preservation. *My grandfather spoke of a time before fences, of endless forests and trees. We've got to have wilderness if want to live free.* I love the freedom of the hills. Today, there is a new threat to freedom and to life. It looks like we must have wilderness to live . . .

Afterword

There are deeper and stronger powers in the wilderness, and the wild places do teach us wisdom when we explore them and listen. I like the lesson that the mountain allows us to be there, and we climb its peaks by *cooperating* with it rather than fighting it. The mountain may seem to pay no attention to us, but we must pay attention to the mountain, the routes, the cracks, the lines. The mountain even draws or pulls, causing us to be *observant* of details and aware of features we don't usually notice. Climate can teach us as well if we observe and cooperate. Mountains and glaciers matter.

> *For most of us sacred mountains are remote from the experience of everyday life. They lie far off in space and time, revered by distant cultures, many of which vanished long ago. Even the peaks that we manage to climb and visit rise on the borders of our lives, removed from the cities and plains where most of us live. What is the value, then, of thinking about them? It is simply this: the contemplation of sacred mountains, with their special power to awaken another, deeper way of experiencing reality, opens us to a sense of the sacred in our homes and communities—a sense that we need to cultivate in order to live in harmony with our environment and with each other. In looking up to the heights and reflecting on the world around them, we discover within ourselves something that enables us to lead deeper and more meaningful lives. (Bernbaum)*

Even though world religious groups support creation care and stewardship of the earth, the greatest opposition to climate change science comes from right-wing evangelical Christians. The reasons for their opposition include: (1) climate and environmental disturbances are signs of the end times; (2) the Creator is in control of the climate and would not allow humans to destroy the earth; (3) government policy and regulations are unnecessary.

Clearly, I believe that these mountain realities are relevant to climate change no matter whether you are living with unacceptable levels of air pollution in a city or having your livelihood threatened by rising sea levels. I believe that we do have a major responsibility to care for the earth which we have been shirking. I believe we must protect our natural environments for our children and our children's children.

Whatever the climate change solutions are, human attitudes and viewpoints must change. That begins with really *observing* what is happening in the Earth's atmosphere affecting local climates and environments with brutal honesty and an open mind. Dealing with and reversing adverse climate effects will mean for individuals and especially for corporations, *cooperating* with nature rather than working against it. *We need to see beyond what we have seen before. It will feel like moving mountains.*

Endnotes

1. "A Century of Ice Retreat on Kilimanjaro." Copernicus Publications. 4 March 2013. https://doi.org/10.5194/7c-7-419.
2. Burns, Cameron M. *Kilimanjaro & East Africa: A Climbing and Trekking Guide*. Seattle: The Mountaineers Books, 2006.
3. Bennett, Jeffrey. *A Global Warming Primer*. Boulder, CO: Big Kid Science, 2016.
4. Burns, Cameron M. *Kilimanjaro & East Africa: A Climbing and Trekking Guide*. Seattle: The Mountaineers Books, 2006.
5. Holmes, Don. *Highpoints of the United States: A Guide to the Fifty State Summits*. Salt Lake City: The University of Utah Press, 2000.
6. Bennett, Jeffrey. *A Global Warming Primer*. Boulder, CO: Big Kid Science, 2016.
7. Bennett, Jeffrey. *A Global Warming Primer*. Boulder, CO: Big Kid Science, 2016.
8. Cox and Fulsaas, Editors. *Mountaineering: The Freedom of the Hills*. Seattle: The Mountaineers Books, 2003.
9. "Climate Change in Kenya." Wikimedia Foundation. 17 September 2024. en.m.wikipedia.org.
10. Burns, Cameron M. *Kilimanjaro & East Africa: A Climbing and Trekking Guide*. Seattle: The Mountaineers Books, 2006.
11. Bennett, Jeffrey. *A Global Warming Primer*. Boulder, CO: Big Kid Science, 2016.
12. "Climate Change in Kenya." Wikimedia Foundation. 17 September 2024. en.m.wikipedia.org.
13. "Climate change and China." Wikimedia Foundation. 11 October 2024. en.m.wikipedia.org.
14. "The Great Wall of China." Wikimedia Foundation. 26 September 2024. en.m.wikipedia.org.
15. "Water Pollution in China." The Borgen Project. 10 March 2018. borgenproject.org.

16. "Global Warming is Worsening China's Pollution Problems." Inside Climate News. 14 August 2019. Insideclimatenews.org.

17. "The Paris Climate Agreement and China." UNFCCC. 30 June 2015. Unfccc.int.

18. Smith, Michael. *An Unsung Hero*. Cork, Ireland: The Collins Press, 2009.

19. "Climate Action." Geological Survey of Ireland. 2021. Gsi.ie.

20. Roach, Gerry. *Colorado's Fourteeners: From Hikes to Climbs*. Golden, CO: Fulcrum Publishing, 1999.

21. "Climate Change Is Changing American Ski Slopes Forever." Conde Nast Traveler. By Jonathon Thompson. 28 December 2020.

22. "Andorra." Wikimedia Foundation. 14 October 2024. En.m.wikipedia.org.

23. "Our History." ASOLO. 2024. Asolo.com.

24. "Climate Change and Andorra." Wikimedia Foundation. 2024. Climateknowledgeportal@worldbank.org.

25. "Climate Change in Alaska." Wikimedia Foundation. 7 October 2022. en.m.wikipedia.org.

26. "Climate Change in Alaska." Wikimedia Foundation. 7 October 2022. en.m.wikipedia.org.

27. Holmes, Don. *Highpoints of the United States: A Guide to the Fifty State Summits*. Salt Lake City: The University of Utah Press, 2000.

28. "History of Alaska." Wikimedia Foundation. 18 September 2024. en.m.wikipedia.org.

29. "What is climate change doing to Wales"? Wikimedia Foundation. 20 September 2019. bbc.co.uk.

30. "Climate Change in the United Kingdom." Wikimedia Foundation. 10 September 2024. en.m.wikipedia.org.

31. "Climate Change in the United Kingdom." Wikimedia Foundation. 10 September 2024. en.m.wikipedia.org.

32. "Climate Change in Argentina." Wikimedia Foundation. 11 September 2024. en.m.wikipedia.org.

33. "The Climate of Aconcagua." MountainIQ. 22 May 2023. www.mountainiq.com.

34. "416 Wildfire Update." *The Denver Post*. 12 June 2018. www.denverpost.com.

35. "Nuclear Power and Climate Change." International Atomic Energy Agency (IAEA). 2019. www.iaea.org.

36. "Climate Change and Australian Bushfires." *BBC*. 4 March 2020. www.bbc.com.

37. "Great Barrier Reef." Wikimedia Foundation. 30 September 2024. en.m.wikipedia.org.

38. "Great Barrier Reef." Wikimedia Foundation. 30 September 2024. en.m.wikipedia.org.

39. Griffiths, Tom. *Slicing The Silence: Voyaging to Antarctica*. Cambridge

Endnotes

and London: Harvard University Press. 2007.

40. "Claude Lorius." Wikimedia Foundation. 13 September 2024. en.m.wikipedia.org.

41. "What the National Climate Report Says about the Pacific Northwest." *The Seattle Times*. 21 November 2023. Seattletimes.com.

Climate Glossary and Definitions

Climate—the conditions of the earth's atmosphere at a particular location over a long period of time.

Weather—the long-term summation of the atmospheric elements and the short-term periods of weather.

Atmospheric elements—solar radiation, temperature, humidity, precipitation, atmospheric pressure, and wind.

Climate change—periodic modification of Earth's climate brought about as a result of changes in the atmosphere as well as interactions between the atmosphere and various other geologic, chemical, biological, and geographic factors within the Earth system.

Timescale—factors that change at very short periods of time: atmospheric chemistry, surface vegetation, and distribution of ocean heat. The position of continents and the location and height of mountain ranges change over very long timescales.

Dynamic motion—climate definitions acknowledge that the weather is always changing, due to instabilities in the atmosphere.

Climate variation—is a redundant expression because climate is always varying.

Global warming—is the most critical issue of climate change in the contemporary world.

Moving Mountains Playlist

Solo Guitarists and Solo Pianists

Aerial Boundaries—Michael Hedges

A Fine Line—Doug Smith

Afton Mountain—Stephen Bennett

Ascent—Joe Bongiorno

Because It's There—Michael Hedges

Breakfast in the Field—Michael Hedges

Cloud Forest—Trace Bundy

County Down from Beyond Nature—Phil Keaggy

Drifting—Andy McKee

Edelweiss (A Mountain Flower)—Ken Bonfield

Everlasting Light—Ryan Farish

Falling Snow—Michael Gulezian

Forest—Mike Howe

Gentle Rain—Brooks Williams

Glorious—David Tolk

Happy Archer—Billy McLaughlin

Ice Fields—Leo Kottke

Jesu, Joy of Man's Desiring—Trace Bundy

Keep It Simple—Tommy Emmanuel

Moving Mountains Playlist

Layover—Michael Hedges
Mile High Country—Michael Gulezian
Moonrise—Trace Bundy
More Travels—Pat Matheny
Night Sky Essays—Liz Story
Night, Part 1: Snow—George Winston
Oregon—Stephen Bennett
Passing Thought—Phil Keaggy
Questions—Tommy Emmanuel
Rainsong—George Winston
Returning—Will Ackerman
Somewhere Within—Joe Bongiorno
The Most Beautiful Sky—Stephen Bennett
The View from Here—Michael Mucklow
Traverse—Trace Bundy
Ursa Major—Michael Hedges
Vincent (Starry Starry Night)—James Bartholomew
Water into Wine—Bruce Cockburn
Wild Mountain Thyme—Ed Gerhard
Wintertide—Don Ross
Your Presence (from Ancient Whisper)—Michael Mucklow
Zuzu's Petals—Will Ackerman

Resources for the Reader

Bass, Dick. *Seven Summits*. New York: Warner Books, 1986.
Bennett, Jeffrey. *Global Warming Primer*. 2016.
Bernbaum, Edwin. *Sacred Mountains of the World*. Cambridge: Cambridge University Press, 2022.
Burns, Cameron. *Kilmanjaro & East Africa: A Climbing and Trekking Guide*. 2006.
Egan, Timothy. *The Worst Hard Time*. Boston: Mariner, 2006.
———. *The Big Burn*. Boston: Mariner, 2010.
———. *Lasso the Wind*. New York: Penguin Random House, 1999.
Holmes, Don. *Highpoints of the United States: A Guide to the Fifty State Summits*. Salt Lake City: University of Utah Press, 1990.
Maslin, Mark. *How to Save Our Planet*. New York: Penguin Life, 2022.
Mountaineering: The Freedom of the Hills. 2003.
Nichols, John. *If Mountains Die*. New York City: Knopf Publishers, 1979.
Roach, Gerry. *Colorado's Fourteeners: From Hikes to Climbs*. 1999.
Robinson, Tri. *Saving God's Green Earth*. Norcross, GA: Ampelon Publishing, 2006.
Smith, Michael. *An Unsung Hero*. 2009.

www.ingramcontent.com/pod-product-compliance
Lightning Source LLC
Chambersburg PA
CBHW052059230426
43662CB00036B/1703